The evolutionary biology of colonizing species

To Louise

The evolutionary biology of colonizing species

PETER A. PARSONS
La Trobe University
Victoria, Australia

CAMBRIDGE UNIVERSITY PRESS

Cambridge
London   New York   New Rochelle
Melbourne   Sydney

CAMBRIDGE UNIVERSITY PRESS
Cambridge, New York, Melbourne, Madrid, Cape Town, Singapore,
São Paulo, Delhi, Dubai, Tokyo

Cambridge University Press
The Edinburgh Building, Cambridge CB2 8RU, UK

Published in the United States of America by Cambridge University Press, New York

www.cambridge.org
Information on this title: www.cambridge.org/9780521272452

First published 1983
This digitally printed version 2010

A catalogue record for this publication is available from the British Library

Library of Congress Cataloguing in Publication data
Parsons, P. A. (Peter Angas)
The evolutionary biology of colonizing species.
Bibliography: p.
Includes index.
1. Evolution.   2. Adaptation (Biology)
3. Habitat (Ecology)   I. Title.   II. Title:
Colonizing species.
QH371.P324   1983   575.1′5   82–19763

ISBN 978-0-521-25247-8 Hardback
ISBN 978-0-521-27245-2 Paperback

Contents

Contents

Preface

In 1964, a symposium on "The Genetics of Colonizing Species" was held at Asilomar, California, which brought together "geneticists, ecologists, taxonomists, and scientists working in some of the more applied aspects of ecology – such as wildlife conservation, weed control, and biological control of insect pests" (Baker and Stebbins, 1965:vii). It was felt that the biology of colonizing species would be more readily understood using a multidisciplinary approach, and, in particular, the conference recognized the need for interaction between population genetics and ecology. Even so, after what was evidently a most stimulating meeting, Mayr (1965) in his summary paper wrote, "The one firm conclusion I can draw is that it is quite impossible to summarize the conference." Since then, there has been progress, but it has been rather sporadic, often arising from investigations having other primary objectives. This book was written in the belief that this situation is about to change. For example, there is renewed interest in attempts at genetic analyses of ecologically significant complex traits, such as life-history characteristics, environmental stresses, and the use of feeding and breeding sites in relation to habitats occupied.

Until recently, genetics, ecology, and the study of behavior have been regarded as separate entities. This view is not realistic in the development of a coherent evolutionary biology of an organism, and we should look forward to a unification of all three disciplines. Nowhere is this more evident than in the study of colonists in their habitats. By using colonizing species as a case study, an object of this book is to demonstrate the importance of such a unified approach. Ultimately, the approach derives from the overriding assumption of the book, which is that the organism is the unit of selection. The organism can be considered as a phenotype with special emphasis on behavior, ecology, metabolism, physiology, morphology, and so forth. However, since this book is on colonizing species and their habitats, the ecobehavioral phenotype including life-history components dominates.

Much of the impetus to write this book came from a workshop held at La Trobe University in March 1981 on "Populations, Habitats and

Colonising Strategies (in Australia)'' (White, Stanley, and Parsons, 1981). At the workshop, ''We hoped that a multidisciplinary approach would generate fresh insights into adaptive strategies which have enabled Australia's plants and animals to spread into such a wide variety of habitats.'' Like the 1964 meeting, the workshop was enormously stimulating for the geneticists, ecologists, anthropologists, and prehistorians who were invited to the meeting. This book can be regarded as a direct outcome of that meeting; the link with the 1964 meeting is shown by its title, *The evolutionary biology of colonizing species*. It must be stressed, however, that I emphasize those organisms with which I have some familiarity. As a result the plant kingdom in particular might be considered by some to be rather neglected, although I have attempted to put plant material into the framework of the book wherever possible.

I am grateful to all participants of the 1981 meeting for many necessary stimulating discussions, especially my coorganizers, Drs. Suzanne M. Stanley and Neville G. White. Both in the organization of the conference and in the preparation of this book (typing, reference work, and diagrams), I am especially grateful to Ms. Marlene Forrester. Others who have assisted in immeasurable ways include Mr. Ary Hoffman, who provided many perceptive criticisms and helped to convince me, together with a most thoughtful letter from Dr. Ernst Mayr of Harvard University, to have a title without the word *strategy*. This is a term that has provided many useful generalizations but is now perhaps leading to some confusion in the evolutionary literature. Dr. Luca Cavalli-Sforza of Stanford University kindly has permitted me to use his calculations on the range expansion of the marine toad, *Bufo marinus* (Figure 10.5). Ms. Jan Clark helped with many of the diagrams. Dr. Lee Ehrman of the State University of New York provided her usual moral support from the other side of the world and together with Mrs. Toni Faucher improved the manuscript in the draft stage.

During early 1982 while writing this book, I had the opportunity to present this material to a few advanced-level students of the Department of Genetics and Human Variation. Their input suggests that this book may be useful for an advanced-level course. To the staff that permitted this opportunity, I give special thanks.

I wish to thank certain mentors who have influenced and helped me over the years. It is impossible to list them all; however, David Catcheside, the late Sir Ronald Fisher, George Owen, Bob Allard, Mel Green, Hampton Carson, Walter Bodmer, John Thoday, Barry Lee, and Bill Palmer must be mentioned. Finally, the enthusiasm of Ledyard Stebbins when he recently heard of this project must be acknowledged.

1

Introduction: Colonists and habitats

An important milestone in the development of ecological population genetics was the symposium sponsored by the International
Union of Biological Sciences at Asilomar, California, in 1964
on *The Genetics of Colonizing Species*. [Stebbins, 1979:40]

This quote is provided to indicate that the most significant previous
major publication on the theme of this book is the volume recording
the activities of the 1964 symposium (Baker and Stebbins, 1965). The
volume from this conference contains many leads. Many are today not
fully developed. A summary of many of the questions posed appears
in the symposium's Introduction (Waddington, 1965) and the Summary
(Mayr, 1965). Before reopening some of these issues, however, it is
important to consider briefly what is meant by the term *colonizing
species*, not as an absolute definition, but to set the scene for the chapters that follow. This is simply because the borderline between a colonizing species and a noncolonist is arbitrary and defined mainly by
accepted usage.

At one extreme, every species is a colonist, since otherwise it would
not occupy any territory. This extremely broad definition is, however,
of little practical use. Often in the literature, considerations are restricted to species invading disturbed and newly created habitats following humans. With increasing urbanization and the destruction of
natural habitats, such species are increasing in importance year by
year. Among these species, many insects, certain rodents, and weedy
plants are notably important (Baker and Stebbins, 1965). Among the
insects, some of the widespread species of the dipteran genus *Drosophila* are important as colonizing species and have been extensively
studied. Dobzhansky (1965) described the colonizing *Drosophila* species as "animal weeds" that occur "in or near human habitations,
gardens, orchards, places of storage of food products, and garbage
dumps and are not found in habitats reasonably remote from such places." These are the types of colonizing species that will mainly be
considered especially in the early parts of this book. In these circumstances, the ecology of habitats colonized will usually be similar to that

1

of the habitats from which the colonists came, that is, genetic changes are likely to be minor except for adaptation at the local level, although in a new zone there may be an ecological opportunity for expansion and genetic diversification. Even so, colonization of this type is unlikely to lead to speciation, since it involves a move into more or less uninhabited territory, which presents no particularly novel environmental conditions apart from relative emptiness at the time of colonization.

In contrast, there is the colonization that occurs when a species shifts into a distinct new ecological niche, whereby a distinct new type of habitat is exploited (Mayr, 1965). Such ecological colonization must be extremely important in evolution. Unfortunately, the direct observation of ecological colonization is difficult, but it has been inferred a number of times (Bush, 1975; Parsons, 1982a). Included in this category may be species that arrived in Hawaii and other oceanic islands before humans affected the environment. In many cases, as in Hawaiian *Drosophila*, major adaptive radiations followed involving range expansions in habitats little disturbed by man (Carson et al., 1970).

It is only beginning to be appreciated that the most extraordinary adaptive radiations occur among parasitic organisms (Price, 1980), which may be regarded as colonists of their hosts. As in the "weedy" species mentioned earlier, no immediate genetic consequences are implied, unless there are changes that are ecologically relatable to variation among hosts. Taking the definition of *parasite* as used by Price (derived from Webster's Third International Dictionary), which is "an organism living in or on another living organism, obtaining from it part or all of its organic nutriment, commonly exhibiting some degree of adaptive structural modification, and causing some degree of real damage to its host," it turns out that the feeding habits of 71.1% of British insects based upon a 1945 checklist can be regarded as parasitic (Table 1.1). The individual of any parasitic species usually gains most food from a single living organism, in contrast with more generalized grazers, browsers, and predators that feed on many organisms during their lifetime and saprophages that feed on dead organic matter. Into this last category come many of the Drosophilidae that feed upon microorganisms in decaying organic matter.

Much of the literature on animal colonists and their habitats and evolutionary adaptations is on *Drosophila*. Even within this genus, one particular type of saprophage category dominates, namely, those species involved in alcoholic fermentation; they are attracted to fermented-fruit baits and are frequently amenable to laboratory culture. This resource-utilization category facilitates colonization in unnatural habitats, in contrast with other more specialist resource categories that preclude such range expansions. Table 1.2 gives a classification of the

Table 1.1 *Feeding habits of British insect species on a checklist of Klóet and Hincks (1945)*

| Order | Predators | Nonparasitic herbivores and carnivores | Parasites | | Saprophages |
			On plants	On animals	
Thysanura					23
Protura					17
Collembola					261
Orthoptera		39			
Psocoptera		70			
Phthiraptera				308	
Odonata	42				
Thysanoptera			183		
Hemiptera	123		283	5	
Homoptera			976		
Megaloptera	4				
Neuroptera	54				
Mecoptera	3				
Lepidoptera			2233		
Coleoptera	215	65	909	18	1637
Hymenoptera	170	241	435	5342	36
Diptera	54	231	922	311	1672
Siphonaptera				47[a]	
Totals	665	646[a]	5941	6031[a]	3646
% of insect fauna	3.9	3.8[a]	35.1	35.6[a]	21.5

[a] These figures differ slightly from those in Price (1977), because the biting flies (Diptera) have been transferred to the category of nonparasitic herbivores and carnivores.
Source: After Price, 1980.

3

Table 1.2. *Drosophila species of Australian rain forests classified according to collection method as an indication of likely breeding sites*

Subgenus	Collection method						
	Fruit	Fruit and mushroom	Mushroom	Bracket fungi and forest fungi	Forest fungi	Flowers	Notes on species collected by sweeping
Drosophila	*sulfurigaster* *pseudotetrachaeta* *persicae* sp. nov. 7	*immigrans*[a] *rubida*					
Sophophora	*simulans*[a] *pseudotakahashii* *ananassae*[a] *pseudoananassae* *ironensis* *bipectinata*	*serrata* *birchii* *denticulata* *dispar*					Three closely related spp.: *pinnitarsus* *scopata* *progastor*

4

Hirtodrosophila					Four closely related spp.:
		polypori	allynensis		zentae
		mycetophaga	macalpinei		junae
		mixtura	hirudo		palumae
		hannae	angusi		durantae
			spp. nov. 3–		spp. nov. 1, 2
			6		Up to 33 spp.
	altera				
	fungi				
	eluta				
	rhipister				
	pictipennis			minimeta	
	mossmana			hibisci	
	oweni				
Scaptodrosophila	bryani				
	fumida				
lativittata[b]					
enigma[b,c]					
specensis					
novoguinensis					

[a] Cosmopolitan species mainly found in newly created habitats following humans.
[b] Species spreading into orchards in Australia.
[c] Species found recently in New Zealand.
Source: Adapted from Parsons, 1981a.

5

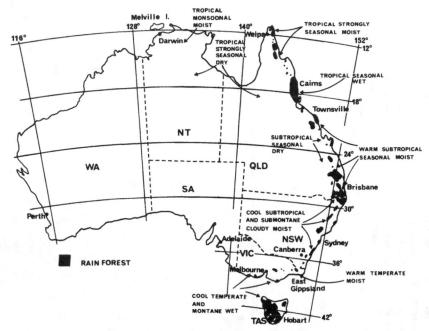

Figure 1.1. Distribution of Australian rain forests. The arrows indicate the approximate center of distribution of each climatic type, but the boundaries of each type overlap because of past climatic changes. Many Australian localities mentioned elsewhere in the text are incorporated in this map. (Source: Modified from Webb and Tracey, 1981.)

Australian *Drosophila* fauna found in rain forests (see Figure 1.1 for the distribution of the major rain forests) according to resource-utilization category of species with a note as to species known to be colonists, that is, cosmopolitan species mainly found in newly created habitats following people. In addition, there are endemic species spreading into orchards in which discernable genetic changes relatable to ecological factors are possible, since orchard resources have only been recently established in Australia. The classification in Figure 1.1 can only be regarded as semiquantitative, because data have been obtained nonrandomly according to species. Even at this stage, it is pertinent to comment that there may be a large fauna of wasp (Hymenoptera) parasites of *Drosophila* that is worthy of detailed study, especially for those *Drosophila* species where sophisticated evolutionary studies are possible. However, consideration of parasites including those of *Drosophila* will be left to Section 11.1, and up to that stage the colonizing animal species that will be discussed come from the remaining categories in Table 1.1.

Since colonists may arise from populations at the margins of the distributions of species, populations from central and marginal habitats must be compared at the genetic level (Carson, 1965). An ecologically marginal habitat can be regarded as one in which physical stresses tend to be both variable and extreme, so that resources tend to be unpredictable and shortlived. Therefore, such habitats are stressful at unpredictable times for organisms occupying them, compared with more centrally located habitats that would usually be more benign. It is important to distinguish marginal habitats defined ecologically, that is, as likely to be stressful from those that are peripheral in the geographical sense. Necessarily, there is often a high correlation between marginal and peripheral habitats, but in areas of high topographic variability, correlations may become somewhat obscure.

A colonizing species must fit into the available ecological niches of its new locality. In some cases, for example, in Australia and New Zealand, there are reasonably accurate estimates of the time periods involved. Phenotypically, a degree of toughness is to be expected, enabling the colonist to cross unsuitable areas or climatic belts before finding a suitable habitat and becoming established. Another important characteristic of virtually all colonizing species is their general flexibility. They must be physiologically tolerant of the conditions of their new habitats, show flexibility in their genetic and mating systems, and not show extreme specialization especially in the utilization of resources. There are a number of features specific to plants that appear in discussions of colonizing species (Mayr, 1963; Baker and Stebbins, 1965; Harper, 1977). These complexities will not be considered in detail, although plant examples will be discussed (in particular, in Section 11.2) to complement animal studies. Although there is an array of breeding systems in colonizing animals, outcrossing is normal in the majority of animals to be discussed. The main exceptions are in Chapter 11, where it will be seen that the parasitic mode of life may encompass several breeding systems, as is true of plants.

Implicit in these comments is the approach of comparing colonizing species with closely related species that are not undergoing range expansions. For example, comparisons of tolerances to environmental extremes will be emphasized in insects, since their powers of dispersal are often limited. Indeed, suitable habitats for insects, such as *Drosophila*, depend intimately upon appropriate physical conditions as a prerequisite for resource exploitation. Their life spans are normally too short for the development of learning ability beyond a rudimentary level as a factor in habitat selection. Higher in the phylogenetic series, however, behavioral factors become progressively more important in insulating organisms from the effects of the microenvironment (Ehrman and Parsons, 1981a). In the final analysis, there is no sharp separation of genetics, ecology, and the study of behavior in the study of animal colonization and of habitats occupied (Mayr, 1965). Behavioral aspects have been least studied, but with increasing studies of the ecobehavioral components that make up habitat selection, this situation is slowly changing especially in vertebrates, in particular, certain rodents. However, since much of the work on animal populations having a significant genetic input has been done on insects, a major thrust of this book is toward the interplay of genetics and ecology. Even so, in later sections, behavioral considerations are emphasized. Discussions of habitats and their heterogeneity are assuming increasing prominence in the literature. This is a trend that is likely to develop further in the future.

The overriding assumption in this book is that natural selection, which guides evolutionary change, acts primarily on phenotypes and, secondarily, on genotypes. Any agent that alters or limits phenotypes, for example, physiological adaptation or learning, may have ultimate evolutionary consequences, since genotypic changes may follow phenotypic changes. It is not an overemphasis to say that the invariable stress is upon selection favoring certain individuals over others through differential reproduction in differing habitats occupied according to their phenotypic characteristics. Indeed, one reason for the rather patchy progress in understanding colonists in their habitats resides in the frequent tendency not to use phenotypic data as the primary data. Proceeding via the phenotype, it is important to ask whether there are genetic architectures that characterize colonists. Although phenotypic comparisions between colonizing species and closely related noncolonists are informative, it will be seen that useful conclusions at the genotypic level are sparser.

2

Genetics and ecology

The subdivision of all natural selection into r selection and K selection is convenient, because it is a fairly natural subdivision, but it is by no means the only possible one. [MacArthur, 1972]

2.1 Analyses of life histories

There are signs that the integration of genetics at the population level with ecology that is needed for a coherent evolutionary biology of populations is beginning (Jain, 1979; Price, 1980; Parsons, 1982a). Population geneticists frequently choose clear-cut genetic markers – chromosomal inversions, recessive lethals, and electrophoretic variants – in order to study population variability. Having assayed variability, a search is made for the ecological significance of the variability patterns, almost as a second-order situation. By contrast, the ecologist takes traits of obvious adaptive significance in determining the distribution and abundance of organisms; the genetic basis of the traits is not often considered. Following Jain (1979), who was mainly considering plants, analyses of adaptation should proceed with (1) comparative studies of variation of several well-chosen traits, (2) the establishment of adaptive roles of these traits, (3) the demonstration of genetic changes under varying environments, and (4) the experimental demonstration of whether the outcomes represent adequate solutions to an adaptive challenge. Until recently, points (2) to (4) were not often considered, but the situation is beginning to change in that regard. This means that inferences are often speculative, since these omissions mean that genetic aspects tend to be given little emphasis – a conclusion that applies to animals and plants alike.

There is a considerable ecological literature on life-history characteristics of populations often expressed in terms of the frequently debated $r-K$ continuum (MacArthur and Wilson, 1967). Under good conditions, the numbers of any population will increase exponentially at a rate determined only by its net reproductive rate (or Malthusian parameter) r (Figure 2.1). Writing N as the population size and ΔN as

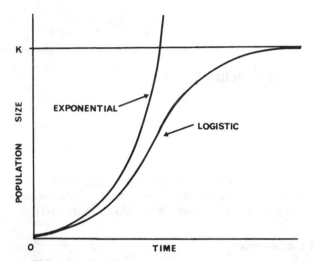

Figure 2.1. Population growth curves. The exponential growth in an unlimited environment depends upon the Malthusian parameter, *r*. In a limited environment, the logistic curve may occur, where the population size approaches *K*, the carrying capacity.

the rate of change of the population size, this rapid increase expressed by

$$\Delta N = rN$$

will not last long, and in due course individuals will begin to react with one another and compete for resources. The increase in numbers will then slow to a point where the population size *N* becomes constant. This is termed the *carrying capacity* (*K*) of the environment. The most commonly used model (the Lotka–Volterra model), taking *K* into account, is written

$$\Delta N = rN \left(\frac{K - N}{K} \right)$$

Because of the term in parentheses, ΔN approaches zero as *N* approaches *K*. The resulting growth curve, which is referred to as the logistic curve, is illustrated in Figure 2.1. The rate of change of *N* depends very much upon the parameter *r*, which is expected to have a high value in colonizing species, that is, the rate of growth per capita in colonizing species tends to be high.

Species are frequently classified as *r* or *K* species (e.g., Southwood, 1977). Those of ephemeral habitats where physical conditions and resource availability are highly variable are typical *r* species. Some synonyms of *r* species listed by Southwood (1977) include fugitive species,

Table 2.1. *The predictions of* r *and* K *selection and the bet-hedging models*

Model	Assumptions	Predictions
Deterministic		
r selection	Exponential population growth; stable age distributions; repeated colonizations or fluctuations in population density	Earlier maturity; more smaller young; larger reproductive effort; shorter life
K selection	Environment stable; population near equilibrium density; logistic population growth; competition important	Later maturity; fewer, larger young; smaller reproductive effort; longer life
Stochastic	Environment fluctuates; population near equilibrium	
	(a) Juvenile mortality or birthrate fluctuates, adult mortality does not	Later maturity; smaller reproductive effort; fewer young
	(b) Adult mortality fluctuates, juvenile mortality or birthrate does not	Earlier maturity; larger reproductive effort; more young

Source: Adapted from Stearns (1977).

vegetative species, opportunists, denizens of temporary habitats, pioneering species, exploitation competitors, fast species, super tramps, small species, and ephemeral species. In large measure, this is a description of colonizing species, where rapid growth and large populations occur at times when the environment is favorable, since at other times they may be threatened with extinction in their marginal habitats. For simplicity, the terms r and K selection will be used in this book, although many refinements are needed before the development of a comprehensive life history theory (Stearns, 1976; Denno and Dingle, 1981); indeed, there are two additional but nonexclusive models (see Tallamy and Denno, 1981) – the bet-hedging model (Stearns, 1976, 1977) and the balanced-mortality model (Price, 1974).

In r–K continuum models, mortality and fecundity schedules do not fluctuate. K selection is said to occur in resource-limited environments favoring the ability to compete and avoid predation, whereas r selection occurs in environments favoring rapid population growth, so leading to groupings of life history characteristics (Table 2.1). Bet-hedging deals specifically with fluctuating (stochastic) mortality schedules. When juvenile mortality fluctuates due to unpredictable environmental conditions, natural selection favors organisms with some traits characteristic of K selection (Table 2.1). In the case of unpredictable adult mortality, natural selection in fluctuating and stable environments is predicted to parallel r and K selection, respectively. The balanced-mortality hypothesis concentrates upon life-history evolution in terms of mortality suffered in a hostile environment. For example, natural selection will balance high mortality levels by increasing egg production, so that the high fecundity of parasites may be an adaptation to a high probability of juvenile mortality.

Arguing from discrepancies that run contrary to all three models in lacebugs, Tallamy and Denno (1981) consider that covarying life-history traits evolve as responses to many interacting selection pressures and that different combinations of traits may be equivalently adaptive in different environments. Furthermore, Parry (1981) considers that recognition of the complex interactions that determine life-history patterns is more likely to provide insights than attempts to fit these patterns into simple but difficult-to-define models, as in Table 2.1. It is not, therefore, surprising that a literature testing models of life-history characteristics is beginning to appear. For example, Barclay and Gregory (1981) tested the predictions of the bet-hedging and r–K continuum models in *Drosophila melanogaster*; they considered that no model predicted experimental outcomes very accurately but that the general predictions of the bet-hedging model seem to be borne out better than the r–K continuum model.

Notwithstanding these difficulties, procedures to estimate r, also called the *instantaneous rate of increase of a population*, may be an important approach to take. For comparative purposes, r_m, the intrinsic rate of increase, is often estimated being r under defined conditions of temperature, moisture, quality of food, and so forth when the quantity of food, space, and other animals of the same kind are kept at an optimum, and organisms of a different kind including predators are excluded. Estimating r_m is not, however, simple being based upon tables of birth and death rates for populations having stable age distributions. There are relatively little published data, although such studies have helped to specify the characteristics of colonizing species (Lewontin, 1965; Safriel and Ritte, 1980), since in the field the rate of increase of some species may typically approach r_m during colonizing episodes when resources are abundant.

The pelagic tunicate, *Thalia democratica*, has one of the fastest rates of population increase of any multicellular animal and is close to that of algae and the other microorganisms upon which it feeds (Heron, 1972a,b). This animal is adapted to rapid continuous growth, with a minimum of expenditure and time on activities not concerned with growth and reproduction. There is a strong inverse correlation of r upon generation time, but r is relatively independent of number of offspring, suggesting strong selection in colonizing species for reduced

Table 2.2 *Mean egg-to-adult viabilities and development times of* r *and* K *selected lines of* Drosophila pseudoobscura

Selected line	Development time (days)		Egg-to-adult viability
	Females	Males	
*r*1	19.53 ± 0.19	20.21 ± 0.18	0.68 ± 0.05
*K*1	20.45 ± 0.31	20.97 ± 0.29	0.82 ± 0.06
*r*3	19.08 ± 0.16	19.71 ± 0.16	0.73 ± 0.07
*K*3	20.66 ± 0.29	20.20 ± 0.27	0.95 ± 0.07

Source: Simplified from Taylor and Condra (1980).

generation time. This result agrees with Cole (1954) and Lewontin (1965), who showed theoretically that the generation time, or age at first birth, is of prime importance in producing a high value of *r*.

Considering various *Drosophila* species, Birch et al. (1963) found large differences in longevity and fecundity between geographical races of *D. serrata*, associated with minimal differences in development time. This suggests that in many populations, development time is very close to its physiological limit. In agreement, Clarke, Maynard Smith, and Sondhi (1961) found a strongly asymmetrical response to artificial selection for development time in *D. subobscura*, with a realized heritability (see Appendix) of 0.06 for downward and 0.19 for upward selection. Even so, some development time differences have been recorded. For example, Giesel (1974) found that populations of *D. melanogaster* from environments with short growing seasons begin reproduction 1 or 2 days earlier than those from which growing seasons are longer. The importance of development time as a fitness trait was emphasized by Giesel (1979) and Giesel and Zettler (1980), who found it to be correlated with adult longevity and larval viability in *D. melanogaster*. However, caution is necessary since these authors used inbred strains and hybrids where fitness components depend upon the number and severity of recessive deleterious alleles that may have been fixed by inbreeding; for alleles of this kind, positive correlations among fitness traits are to be expected (Rose and Charlesworth, 1981). In *D. pseudoobscura*, Taylor and Condra (1980) established population cages and selected some for rapid generation cycles under uncrowded conditions (*r* selection), and others for abilities to withstand crowding (*K* selection). After 10 months, corresponding to 17 generations of *r* selection, it was found that the *r*-selected populations developed somewhat more rapidly and the *K*-selected populations survived better (Table 2.2). There were no differences opposite to those predicted by the *r*–*K* continuum model for any trait, but there were no differences

Table 2.3. *Ecological characteristics of the*
Queensland fruit fly, Dacus tryoni *(r strategist)*

1. Forms temporary populations
2. Multivoltine (short, overlapping generations)
3. High capacity for increase
     Rapid development (4–6 weeks egg to egg)
     High fecundity (perhaps 100 eggs/female/week)
     High survival in suitable fruit
     Long-lived adults (more than 7 months observed in field)
4. High capacity for dispersal
     Maximum recorded movement, 24 km
     Strong juvenile dispersal
5. Polyphagous (more than 150 known hosts)
6. Diapause unknown
7. Few effective parasites
8. Odor oriented (sex pheromones, protein foods)

Source: Adapted from Bateman (1977).

for several traits about which the model is explicit in predicting differences. These include intrinsic rate of increase, carrying capacity, body size, fecundity, and timing of reproduction. Overall, the main predictions of the model were weakly met; however, development time may be somewhat more variable than envisaged some years ago.

The Queensland fruit fly, *Dacus tryoni*, occurs from northern Queensland to southeastern Victoria over a climatic range from tropical to temperate. *D. tryoni* is found where the distribution of an essential resource, fruit for oviposition and larval food, is discontinuous both in space and time. In the rain forests where *D. tryoni* evolved, fruit tends to occur in aggregations that develop, remain for a time, then disappear. This is followed by other aggregations in the same general area, which repeat the same cycle (Bateman, 1977). The unpredictability involved argues for an *r*-selected species. Bateman (1977) described the life history of *D. tryoni* (see Table 2.3):

> *D. tryoni*'s life history strategy fits it extremely well to the transient nature of its larval food supply. Its high mobility enables it to discover areas where fruit is developing, then its high capacity for increase and rapid overlapping generations enable it to exploit to the full the resources that become available in that area. As it does so it produces large numbers of juvenile adults, a high proportion of which disperse to seek out other areas where fruit may be developing. When the exploitation of the first area is almost complete, the adults still present there also enter a dispersive and searching phase, to locate other aggregations of fruit. They are assisted in this by their considerable longevity, their high mobility, and their poly-

Table 2.4. *Mean lethal dose (LT$_{50}$) hours for adults of four strains of* Dacus tryoni

	Latitude (°S)	Temperature (°C)	
		0	37
Cairns	17	36.9	25.6
Brisbane	27	37.4	30.5
Sydney	34	40.1	32.4
East Gippsland	38	40.4	40.3

Source: Data from M. A. Bateman in Lewontin and Birch (1966).

phagy or host-flexibility. For the Queensland fruit fly, any one of about 150 different hosts will suffice.

Among other things, this provides an excellent insight into the reasons why *D. tryoni* is such a remarkably efficient pest, since in a climatically favorable region, it will almost certainly locate every orchard or even isolated tree where fruit is developing. All of the ecological characteristics in Table 2.3 contribute to this efficiency, but the key character is mobility (Bateman, 1977).

Considering four populations (Bateman, 1967) from Cairns (C), Brisbane (B), Sydney (S), and East Gippsland (Figure 1.1), the r_m ranking was C > B > S at 30°C, which is an agreement with the climatic gradient, whereas at the cooler 20°C, the reverse S > B > C occurred. The populations could not be differentiated at the intermediate temperature of 25°C. Therefore, the results at extreme temperatures show differences between the three populations in the expected directions. The southern East Gippsland population appeared better adapted to both higher and lower temperatures than the more northern populations, which is to be expected since temperatures that are both higher and lower than those to the north occur in this habitat. In parallel, Bateman (Table 2.4) (see Lewontin and Birch, 1966) found that tolerance of extremes of temperature increased from the tropics to the temperate zone, confirming the extreme nature of the East Gippsland population in this respect. Indeed, it is only in the last 100 years that *D. tryoni* has expanded its range so far south (and to higher altitudes in southern Queensland). Apparently, *D. tryoni* was originally dependent upon the fruits of tropical rain forests, but with the introduction of cultivated fruits, it spread to orchards in Queensland close to its endemic home. Subsequently, it spread to orchards in climatically more extreme habitats as a result of physiological adaptation to extreme temperatures. More detailed analyses show that the major differences among populations (at a given temperature) concern the proportion of females fertilized in experimental cages. There were also significant differences in adult longevities and survival rates of immature stages, but not in development time, a result fairly consistent with the afore-mentioned *Drosophila* data.

The important conclusion from this *Dacus* study is that there are climatic races that can be related to temperature in the wild and can be detected by exposure to extreme temperatures. The same is true of *Drosophila melanogaster* (Parsons, 1980a). An Australia-wide analysis shows that extremes of temperature play a significant role in biogeography and, hence, the distribution of organisms (Nix and Kalma, 1972). The qualitative agreement of the extremely tediously obtained r_m results and the relatively simple temperature stress experiments suggest that an important approach for the study of marginal populations and colonizing species is the laboratory study of the effects of biologically realistic environmental stresses, principally temperature extremes and desiccation.

Dingle (1974) studied three species of cotton stainer bug, *Dysdercus*, from East Africa. *D. fasciatus* feeds on the fallen fruits of the baobab tree (*Adansonia*), which is a seasonally rich resource. Resource exploitation is maximized by early reproduction and fecundity; an estimate of $r = 0.0939$ per individual per day. *D. nigrofasciatus* feeds on a variety of generally available annual and perennial Malvales that are never as abundant as baobab fruits in season giving an estimate of $r = 0.0878$. In addition to Malvales, *D. superstitiosus* feeds upon a variety of other plant species, giving an estimate of $r = 0.0616$. Hence, reproductive patterns are consistent with resource availability fluctuations in these three habitat categories. From later work of Dingle (1981) and his colleagues, r values for two migratory *Dysdercus* species came to 0.094 and 0.106, and for two nonmigratory species, 0.062 and 0.066. Migration in insects is frequently associated with the colonization of new habitats being empty universes for the species in question. Selection under these conditions would favor those individuals that can produce descendents most rapidly and in the greatest numbers, that is, r-selection traits would be favored as the aforementioned figures indicate.

There is also a positive association of body (and egg) size with migratory capacity. The largest species of *Dysdercus* in Africa and America are both *specialist* migrant colonizers of oil-rich seeds of trees in the family Malvales, which are far apart in space and time (Derr, 1980). Large body size confers an advantage when resources are temporary, because it allows increased rate of egg output and increases the chances of surviving long periods of migration and diapause. Within and between nine *Dysdercus* species, larger females produce more eggs per clutch, and at a faster rate, than small females. Intermediate-sized species of *Dysdercus* are *generalist* migrant colonizers of a wide array of

Table 2.5 *Comparison of* Tribolium castaneum *and* T. confusum *for certain correlates of the rate of increase,* r, *under standard laboratory conditions of 29°C, 70% RH, together with emigration rates*

	T. castaneum	T. confusum
Development time[a] (egg to adult), days	29.5	31
Onset of reproduction[a] (time after eclosion), days	4	5–6
Fecundity per day[a] (first 3 months of adult life), eggs	9–24	7–11
Average female life span,[a] months	7–8	10–12
r per head/day[a]	0.128	0.100
Mean % emigration based on cohorts of 100 beetles[b]		
24 hr	72	15.5
360 hr	99.7	66.5

[a] Data tabulated by Ziegler (1976).
[b] Data collected by Ziegler (1976).

seeds from annuals and shrubs that are more evenly distributed than trees (Derr, Alden, and Dingle, 1981). These insects need not fly so far nor wait so long between seed crops as the arboreal specialists, but explosive reproduction is still advantageous. The smallest species, which has a noncolonizing life history, reproduces on the seeds of herbaceous weedy annuals. Females reduce the risk of individual mortality by minimizing egg volume and scattering egg clutches across low-density resources.

From a review of the literature on the beetles *Tribolium castaneum* and *T. confusum*, Ziegler (1976) concluded that both are colonists of temporary habitats, but that *T. castaneum* is a primary and *T. confusum* a secondary colonist. Comparative values of *r* and several other life-history traits are in agreement (Table 2.5). It would, therefore, be predicted that *T. castaneum* would show higher intrinsic levels of emigration than *T. confusum*. Ziegler demonstrated this in the laboratory by showing that *T. castaneum* emigrated almost to the last individual in all experiments but that *T. confusum* tended to emigrate more slowly and ceased emigration at low population densities (Table 2.5).

In the mosquito, *Aedes aegypti*, Crovello and Hacker (1972) studied r_m and R_0 (the net reproductive rate). Feral strains were found to have smaller r_m and R_0 values than urban strains, showing that they have differing evolutionary responses to their habitats. The differences between the two sets of strains suggest that the risk of an individual *Aedes aegypti* surviving to reproduction is higher in the ecologically marginal urban than in the feral environment. Apparently, feral strains devote less energy to the production of eggs than do urban strains. That more sophisticated genetic studies are possible with *A. aegypti* is shown by observations of Machado (reported in Crovello and Hacker, 1972) on controlled crosses and backcrosses of inbred strains, indicating that mean life time and r_m are genetically controlled. There is a parallel in plants, since ecologically marginal populations of the sweet vernal grass, *Anthoxanthum odoratum*, have higher turnover rates and lower individual life expectancies than central populations (Grant and Antonovics, 1978).

Istock, Wasserman, and Zimmer (1975) studied the ecology and evolution of the pitcher plant mosquito, *Wyeomyia smithii*. This species is an obligate colonizer, following the seasons and the years of production of new water-filled leaves of the pitcher plant, *Sarracenia purpurea*. Values of *r* were about one-tenth to one-hundredth of those for the blood-feeding mosquito, *Aedes aegypti*. They present arguments that the evolution of autogeny (not feeding on blood) from a previously anautogenous state is associated with a decline in *r*. In contrast with the homogeneity of development times suggested by Lewontin (1965) for colonizing species, the larval population contains a mixture of more rapidly and slowly developing phenotypes, corresponding to a mixture of *r* and *K* phenotypes. This approximates Roughgarden's (1971) model for density-dependent selection and maintenance of a polymorphism by seasonality. The heritability of development time was estimated to be about 0.3; there were considerable fitness variations in natural populations ranging from fast-developing, diapause-resisting, multivoltine, and high-*r* genotypes, to slow-developing, diapause-prone, probably univoltine, and low-r genotypes. Artificial selection for fast development time reduced the heritability to zero in 15 generations. Thus, natural selection must be maintaining the variation in development time in the natural population (Istock, Zisfein, and Vavra, 1976). In fact, natural populations have been shown to experience an alternation of density-dependent (in spring and late fall) and density-independent (summer and early fall) selection in the same season as a consequence of available resources, showing that the position of the insect on the *r–K* continuum varies on a seasonal basis (Istock, Vavra, and Zimmer, 1976). Hence, colonization of a specialized monophagous resource can lead to adaptations, whereby that resource is tracked more closely than the generalist colonizing species of disturbed habitats mainly under discussion, where *r* selection and rapid and uniform development times are expected.

This comparative approach, therefore, demonstrates that estimates of r_m are relatable to habitats defined principally by climate and resources (Andrewartha and Birch, 1954). However, the phenotype represented by these estimates is too complex for anything more than relatively simple genetic analyses. In other words, a prerequisite for more sophisticated genetic analyses are phenotypes relatable to the predictions of the $r-K$ continuum and related models for the evolution of life histories. Tolerance to extreme physical environments and development time appear to be possibilities and will be considered further with other traits, including measures of resource utilization. Climate or physical factors, and resources or biotic factors, are not entirely separable, since the nature and availability of resources depends at least partly upon climate. In the first instance, however, climate and resources will be considered separately, but when habitats are increasingly considered, the separation will be seen to be rather artificial. With regard to development time, it can be concluded that it is relatively invariant in a given environment (say temperature), because of natural selection for rapid development time, but that contrary to earlier views, there are occasional variations in the trait that can potentially be related to habitat. Indeed, it is now becoming clear that genetic differences for fitness characters, either expressed or potential, are common among natural populations and should relate to past selective regimes (Istock, 1981), since geographical differentiation within a species implies past directional selection.

2.2 Summary

For a coherent evolutionary biology of populations, an integration of genetics at the level of populations and ecology is needed. There is a considerable literature on the life-history characteristics of populations expressed in terms of three main models, but discrepancies from all three are appearing. This is because covarying life history traits evolve in response to many interacting selection pressures, whereby different combinations of traits may be equivalently adaptive in similar environments.

For simplicity, however, in this book, discussions will be based upon the $r-K$ (Malthusian parameter, carrying capacity) continuum of characteristics. Colonizing species are characterized by high r values, whereby rapid growth and large population sizes occur at times when the environment is favorable. However, the phenotype represented by estimates of r is too complex to handle for anything more than relatively simple genetic analyses. Therefore, a prerequisite for sophisticated genetic analyses are phenotypes relatable to the predictions of the $r-K$ continuum. These include tolerance to extreme physical environments, development time, and certain aspects of resource utilization.

3

Physical conditions, resources, and ecological phenotypes

Climate plays an important role in determining the average numbers of a species, and periodical seasons of extreme cold or drought seem to be the most effective of all checks. [Darwin, 1859, Chap. III]

Waddington: Has anyone done experiments to try to determine the temperature tolerances or range of foods which these things will accept? Taking your widespread species, have they got a wider temperature tolerance or do they get slowed up less by lower temperature etc.? [Discussion at end of Dobzhansky (1965:549), who was writing on *Drosophila* species.]

3.1 Hard (and soft) selection

The physical conditions tolerated and the resources utilized by organisms play a vital role in determining their distribution and abundance. Collectively, these traits can be called *ecological phenotypes*, implying a relationship with the $r-K$ continuum. Extremes of physical conditions cause death after various periods of time irrespective of the density or the frequency of organisms, that is, populations are subject to *hard selection* in the sense of Wallace (1981).

The difference between hard and soft selection is well illustrated using an example presented by Wallace (1981) on two methods whereby a plant breeder can obtain seed while carrying out a program of artificial selection for increased yield. The breeder plants his experimental seed in a randomized series of test plots (Figure 3.1). At maturity, there are two procedures whereby selected heads may be chosen, which arise because of variation of soil conditions, other environmental variations among plots, or genotype differences. Either the breeder may decide to retain all heads containing 60 kernels or more which means seeds from four of the eight experimental plots only (Figure 3.1, left) or alternatively the breeder selects a small sample of heads from all eight experimental plots (Figure 3.1, right). In the first case, a careless breeder could set his sights too high, and as a consequence, discard

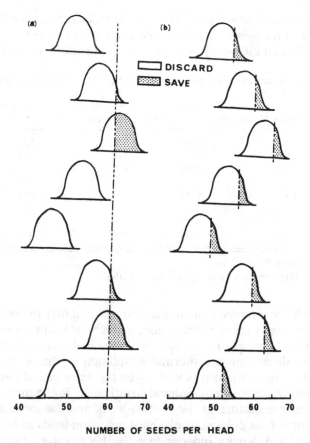

NUMBER OF SEEDS PER HEAD

Figure 3.1. Two artificial selection schemes that illustrate the difference between hard and soft selection. In a, the small-grain breeder saves all heads bearing 60 or more seeds for planting and future selection; some experimental plots are entirely discarded under this scheme. In b, the breeder first samples a few heads from plants of each experimental plot, determines the statistical distribution of seeds per head for each plot, and then harvests what is estimated to be the best 2% to 3% of all heads of each plot for planting and further selection. In this scheme, a few heads are saved from each plot regardless of its average number of seeds per head. Soft selection resembles the second scheme, hard selection the first. (Source: Wallace, 1981.)

all of his material. This procedure imposes *hard selection* on the population of wheat such that there are no survivors. In the second case, it is not possible to discard all the material; indeed, the fate of an individual head of wheat depends upon the phenotypes and genotypes of its neighbors—a procedure described by Wallace as *soft selection*.

At the boundaries of the distributions of species, it can be envisaged

Table 3.1. *Egg-to-adult survival of* Drosophila melanogaster *and* D. simulans *over two temperature ranges, calculated as the areas under the distribution curves of percent emergence plotted against temperature*

Source	10°–31°C		18°–25°C	
	D. melanogaster	*D. simulans*	*D. melanogaster*	*D. simulans*
Lebanon	1379	1161	529	520
Luxor, Egypt[a]	1287	1078	488	490
Alexandria, Egypt	1213	1089	459	484
Wadi-el-Natroun, Egypt	1097	1070	413	478
Uganda	1027	1053	396	473

[a] *D. simulans* from Beni-Swef, Egypt.
Source: Wallace (1981) based on Tantawy and Mallah (1961).

that there would be phases (which may be quite short) of extreme environmental stress or resouce limitation, whereby selection is severe leading to dramatic changes in the distribution and abundance of species. Indeed, in discussing the thermal adaptation of the Australian flora, Nix (1981) stressed that it is very often the extremes of temperature that are of greatest biogeographical significance, for example, the mean maximum temperature of the hottest week and the mean minimum temperature of the coldest week. Such selection tends to be hard and is frequency and density independent. In this chapter, the distribution of *Drosophila* will be considered at the interspecific level, and then in the next chapter attention will be turned to variation at the intraspecific level. Given the dependence of *Drosophila* upon plants, Nix's (1981) conclusion is an important lead to follow.

Detailed comparisons of closely related sympatric species are likely to be especially important in understanding distributional limits. For example, Tantawy and Mallah (1961) measured the percent emergence (egg-to-adult survival) of the sibling species *D. melanogaster* and *D. simulans* at different temperatures. Table 3.1 was derived by Wallace (1981) by taking the areas under the smooth curves fitted to their data. Summed over the entire temperature range studied, 10° to 31°C, *D. melanogaster* has the greatest survival, whereas within the more restricted central portion of the temperature range, *D. simulans* exhibits the highest proportion of emergence. These results imply that the relative fitness of *D. melanogaster* would exceed that of *D. simulans* in a climate that frequently includes temperature extremes but that the reverse would be true under a more moderate and stable temperature regime. Hosgood and Parsons (1966) set up four strains of *D. melanogaster* and three sympatric strains of *D. simulans*, each derived from single inseminated females collected in the wild, at 29.5°, 27.5°, 25°, 20°, and 15°C in the laboratory. After five generations, all strains of *D. melanogaster* were living at all temperatures, whereas for *D. simulans* the three strains at 20°C were living but only one at 25°C (which in fact survived for 24 generations). At 29.5° and 15°C, all the *D. simulans* strains had died out by the second generation, and at 27.5°C, by the third generation. Thus, *D. simulans* is much more restricted in its tolerance to diverse temperatures than *D. melanogaster*, and this agrees with U.S. distribution data reported by Wallace (1968).

McKenzie and Parsons (1974a) have published suggestive evidence for this contrast from field collections of the two species over 3 years in Melbourne, Australia (Figure 3.2a,b); the regression coefficients for the *D. melanogaster*:*D. simulans* ratios are 0.07 and 0.24 when plotted against temperature and temperature fluctuation, respectively, showing the advantage of *D. melanogaster* when temperature stresses are highly variable and often extreme. Even though this appears to be a good example of hard selection such that density and frequency independence apparently prevail, it is not absolutely possible to guarantee this, simply because other events may intervene in determining those organisms that live, die, or become sterile. However, in the marginal environments experienced by potential colonists, the climatic variability experienced would almost guarantee the predominant importance of hard selection.

3.2 Physical stresses

The genus *Drosophila* is believed to have evolved in the tropics (Throckmorton, 1975) and spread to the temperate zone. This range expansion implies a climatic change from uniformly warm humidity to a seasonal pattern of cool, wet winters and warm, dry summers with temperature extremes frequently exceeding those of the tropics. During this range expansion, species should become more tolerant of the climatic extremes of nontropical habitats (Dobzhansky, 1950). Comparison of *Drosophila* species from different habitats (Figure 3.3) supports this prediction, since widespread species are generally more resistant to high-temperature/desiccation and cold-temperature stresses than those restricted to tropical rain forests (Levins, 1969; Parsons and McDonald, 1978; Stanley et al., 1980; Parsons, 1981a).

The ability to invade new areas also depends upon ecological versatility for breeding sites. For example, populations of *D. melanogaster* appear to be both ecologically versatile with respect to the diversity

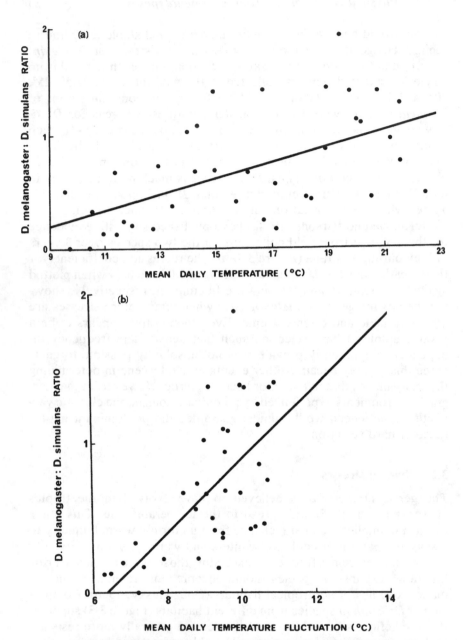

Figure 3.2. (a) Regression of the *Drosophila melanogaster*-to-*D. simulans* ratio (based on combined sexes data) on mean monthly temperature for Melbourne. Regression equation: $y = 0.07x - 0.41$. (Source: McKenzie and Parsons, 1974a.) (b) Regression of the *D. melanogaster* to *D. simulans* ratio (based on combined sexes data) on mean temperature fluctuation for Melbourne. Regression equation: $y = 0.24x - 1.52$. (Source: McKenzie and Parsons, 1974a.)

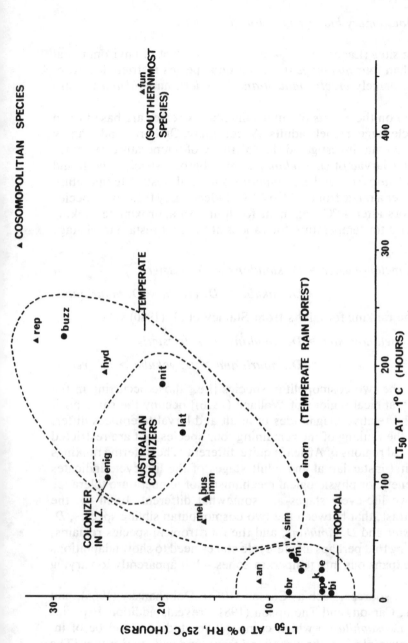

Figure 3.3. LT$_{50}$ values expressed as a plot of the number of hours at which 50% of flies (sexes combined) died of −1°C stress vs. 0% RH 25°C stress for various *Drosophila* species: an, *ananassae*; bi, *bipectinata*; br, *birchii*; bus, *busckii*; buzz, *buzzatii*; e, *erecta*[a]; enig, *enigma*; fun, *funebris*; hyd, *hydei*; imm, *immigrans*; inorn, *inornata*; k, *kikkawai*; lat, *lativittata*; m, *mauritiana*[a]; mel, *melanogaster*[a]; nit, *nitidithorax*; p, *paulistorum*; rep, *repleta*; sim, *simulans*[a]; t, *teissieri*[a]; y, *yakuba*[a]. Superscript[a] indicates *melanogaster* subgroup species. (Source: Adapted from Stanley et al., 1980, and Parsons, 1982a.)

of breeding sites (Lachaise, 1974) and more tolerant of environmental extremes than four *melanogaster* subgroup species restricted to tropical Africa, namely *erecta, mauritiana, teissieri,* and *yakuba* (Figure 3.3).

Most data on the effects of environmental extremes are based upon one life-cycle stage, namely adults. Accordingly, Orsborne and Stanley (unpublished data) investigated the tolerance of temperature extremes of first-instar larvae of six *melanogaster* subgroup species and found that *D. melanogaster* and *D. simulans* survived best at temperature extremes over a range from − 1° to 40°C. Indeed, only these two species had survivors after 40°C treatment for 6 hr. An approximate ranking of the species for temperature tolerances at the first-instar larval stage was

D. melanogaster ≅ *D. simulans* > *D. teissieri*

≅ *D. yakuba* ≅ *D. erecta* ≫ *D. mauritiana,*

whereas the ranking for adults from Stanley et al. (1980) was

D. melanogaster ≫ *D. simulans* > *D. teissieri*

≅ *D. mauritiana* > *D. yakuba* > *D. erecta*

Therefore, the two cosmopolitan species [i.e., those occurring in the six biogeographical regions of Wallace (1876)] occupy the same position, but the relative magnitudes of adult and larval tolerances differ. However, the ranking of the remaining four species that are restricted to the tropical regions of Africa is quite different. The differing rankings between first-instar larval and adult stages of the life cycle indicates that the genetic or physiological mechanisms of temperature tolerance at these two life-cycle stages are somewhat different. Even so, the major contrast, that between the two cosmopolitan sibling species, *D. melanogaster* and *D. simulans*, and the four tropical species remains, so suggesting that perhaps all life-cycle stages need to show adaptations for range expansions into temperate zones – but apparently to varying degrees.

Perusal of sources, such as Carson (1965), Dobzhansky (1965), and Ashburner, Carson, and Thompson (1981), reveals additional species groups of *Drosophila* on which comparative studies would be of interest, for example, the *obscura, immigrans,* and *willistoni* groups. The *willistoni* group consists of five species; four, *D. willistoni, D. paulistorum, D. equinoxalis,* and *D. tropicalis,* are widely distributed in Central America and in tropical and subtropical regions of South America, whereas a fifth, *D. insularis,* comes from certain Caribbean islands (Dobzhansky and Powell, 1975a). The distribution of *D. willistoni* is the widest, ranging from southern Florida to Buenos Aires, Argentina, and in the more central part of its range, the four widely distributed species are frequently sympatric. Compared with the relatively wide-spread *D. paulistorum*, Dobzhansky (1965) regards *D. willistoni* as the most successful colonizer seeming to be "less closely adapted to the particular local environments of the territories it inhabits."

With the exception of *D. ananassae* that is cold sensitive and restricted to the tropics, the remaining cosmopolitan *Drosophila* species have reasonably high levels of resistance to physical stresses as adults (Figure 3.3). The cosmopolitans, therefore, provide an indication of stress levels flies must tolerate in order to spread into the temperate zone. Indeed, based upon the argument that distributional differences between the cosmopolitan species are due to differences in their ability to tolerate environmental extremes, the following ranking was predicted from published distribution data on these species (Parsons and Stanley, 1981) for tolerances to cold: *funebris > immigrans > melanogaster > simulans > ananassae*. The available data were inadequate to place *D. hydei*, *D. repleta*, and *D. busckii*, but they fall between the extremes and are certainly temperate-zone species. In confirmation of this ranking, the following laboratory cold tolerances (Figure 3.3) were found: *funebris > repleta > hydei > immigrans > busckii > melanogaster > simulans > ananassae*. The extreme cold tolerance of *D. funebris* is of interest, since it is common at high northern latitudes, whereas in the southern hemisphere, it is the only *Drosophila* species recorded from southern New Zealand (Harrison, 1959), and in Chile in South America it is the only *Drosophila* species known south of latitude 44°S (Brncic and Dobzhansky, 1957). At the interspecific level, therefore, temperature is clearly among the most important of factors determining distribution patterns.

Among the endemic fauna of Australia, three species, *D. enigma*, *D. lativittata*, and *D. nitidithorax*, are attracted to fermented-fruit baits and are spreading into orchards from natural habitats, and so are in the actual process of colonization. They plot into the temperate-zone cosmopolitan species range and are more resistant to stresses than *D. melanogaster* (Figure 3.3). The colonization potential of *D. enigma* is further confirmed by its recent discovery on the North Island of New Zealand (Parsons, 1980b). It can be argued that apart from these three colonizing *coracina* group species, other Australian endemics are unlikely to be found in New Zealand due to their narrow resource specificity (Table 1.2) and/or sensitivity to environmental stresses. A fourth *coracina* group species, *D. specensis*, is restricted to rain forests and shows no colonizing tendencies; qualitative observations of such field and laboratory flies that have been available indicate a sensitivity to desiccation toward the tropical species in Figure 3.3.

D. inornata is a temperate-zone Australia endemic species showing a marginal tendency to be in urban regions. It is not often attracted to baits and is mainly collected in permanently moist temperate rain forests, so it should be relatively sensitive to desiccation as found (Figure 3.3). Even so, it is more resistant to desiccation than a number of other endemic species that are completely restricted to rain forests (Parsons, 1981a). The desiccation resistance of *D. inornata* appears consistent with its presence in urban collections in winter and early spring only but would be restrictive for wider range expansions. Colonizing *Drosophila* species must, therefore, be sufficiently tolerant of physical stresses in new habitats for resource exploitation to occur.

3.3   Resource utilization

Drosophila species attracted to fermented-fruit baits all apparently utilize ethanol as a resource in vapor form. The effects of metabolic vapors can be assessed by a procedure (Starmer, Heed, and Rockwood-Sluss, 1977; Parsons, Stanley, and Spence, 1979) whereby adults are exposed to the metabolic vapor and water vapor (Figure 3.4). From the number of flies alive after various time intervals, LT_{50}s expressed as the number of hours at which 50% of the flies died can be calculated by linear interpolation. From these LT_{50}s, two measures are particularly useful for describing the effects of the vapors, namely, the concentration at which LT_{50} maximum/LT_{50} control occurs and the threshold concentration where LT_{50}/LT_{50} control $= 1$. When this ratio >1, the vapor is utilized as a resource, and when <1, it is a stress. Resource-utilization curves expressed as LT_{50}s over ethanol vapor divided by control LT_{50}s, are presented for a number of species in Figure 3.5. They cluster into three groups for ethanol resource-utilization thresholds and LT_{50} maximum/LT_{50} control values:

1.   *D. melanogaster,*
2.   *D. lativittata, D. enigma,* and *D. nitidithorax,*
3.   *D. inornata* and *D. hibisci.*

D. melanogaster (1) has an extremely high tolerance compared with the three fruit-baitable *coracina* group species (2), which in turn are able to utilize ethanol more effectively than *D. inornata* and *D. hibisci* (3). *D. hibisci*, which is restricted to endemic *Hibiscus* flowers and so is not a colonizing species, does not apparently utilize ethanol. On the other hand, the low level of ethanol utilization of *D. inornata* may be a factor in its somewhat widespread nature, including its marginal tendency to spread into urban regions.

Species not attracted to fermented-fruit baits, therefore, appear not to have the biochemical phenotype for major range expansions. In addition, *D. hibisci* and *D. inornata* have only "low-activity" alcohol dehydrogenase (*Adh*) alleles; indeed, electrophoretically they are ef-

ETHANOL TESTING APPARATUS

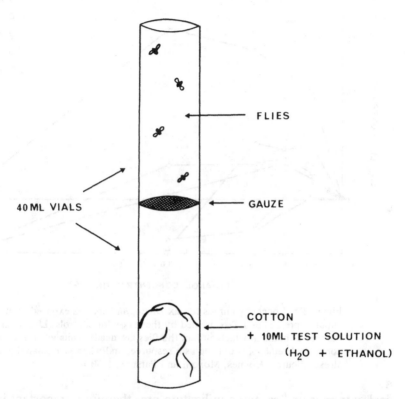

Figure 3.4. Procedure for assaying the effects of metabolic vapors. Flies are admitted to the upper vial, which is separated by Terylene (Dacron) cloth from the lower vial in which a test solution is added. The flies in the upper vial are, therefore, exposed to a constant amount of atmospheric ethanol (water vapor in the controls) in equilibrium with the liquid phase in the lower vial containing the cotton. (Source: After Starmer, Heed, and Rockwood-Sluss, 1977; Parsons, Stanley, and Spence, 1979.)

fectively *Adh*-null (Table 3.2). The *coracina* group species are polymorphic for "high-activity" and "low-activity" alleles at the *Adh* locus, which means that rapid changes in theory could occur during a colonization phase compared with the homozygous low-activity species. Alcohol dehydrogenase (ADH) activity follows the ethanol threshold sequence of $1 > 2 > 3$, so that the species show an association of *Adh* genotype, ADH activity, laboratory ethanol utilization, and resources utilized (Holmes, Moxon, and Parsons, 1980).

Comparative studies of closely related species for tolerance to phys-

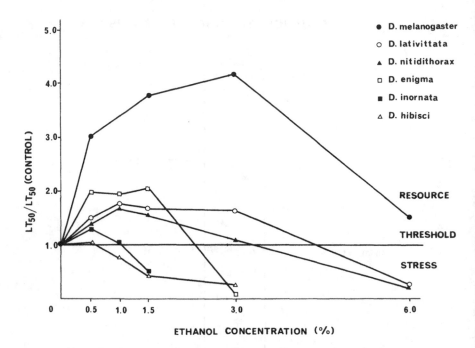

Figure 3.5. Longevity curves for six *Drosophila* species exposed to ethanol vapor expressed as LT_{50}s divided by the LT_{50}s for controls (LT_{50} being the number of hours at which 50% of the flies are dead). Thus when LT_{50}/LT_{50} control >1, ethanol is utilized as a resource, and when < 1 ethanol is a stress. (Source: Holmes, Moxon, and Parsons, 1980.)

ical extremes and resource utilization are, therefore, important in determining potential for spread. Insights into the adaptations of generalist colonizing species should then emerge. In the subfamily Dacinae of Tephritid fruit flies that includes the genus *Dacus*, Birch (1965) lists 54 species that are native to Australia, of which most have relatively few hosts or are monophagous. Species such as *D. tryoni*, on the other hand, have many hosts – both native and imported. Since *D. tryoni* is a colonizing pest species, comparative studies as previously discussed would be of interest in addition to those considered in Table 2.3.

3.4   Resource utilization in three sympatric species

Arguing from Figure 3.5, those *Drosophila* species attracted to fermenting fruits would be expected to use ethanol as a resource. Until recently, ethanol was regarded as rather toxic to *D. simulans*, compared with *D. melanogaster*. Since the two sibling species are frequently sympatric, a more plausible hypothesis is that both species

Table 3.2. *Alcohol dehydrogenase-specific activities and allelic frequencies among endemic Australian Drosophila species*

Species	Subgenus	Collection method	Sex	Adh-specific activity[a]	Allelic frequencies[b]
melanogaster (Townsville)	Sophophora	Fermented-fruit baits	♀	7.3 ± 0.4	n.d.
			♂	8.8 ± 0.7	n.d.
lativittata (Fairfield)	Scaptodrosophila	Fermented-fruit baits	♀	2.6 ± 0.5	a 0.57, b 0.43
			♂	2.5 ± 0.3	
nitidithorax (southwestern Australia)	Scaptodrosophila	Fermented-fruit baits	♀	3.4 ± 0.9	a 0.63, b 0.37
			♂	2.9 ± 0.3	
inornata (Victoria)	Scaptodrosophila	Sweeping forest habitats	♀	1.1 ± 0.3	a 0.00, b 1.0
			♂	0.9 ± 0.2	
collessi[c] (Victoria)	Scaptodrosophila	Sweeping forest habitats	♀	0.8 ± 0.2	a 0.00, b 1.0
			♂	0.4 ± 0.1	
hibisci (Sydney)	Scaptodrosophila	Sweeping Hibiscus flowers	♀	0.7 ± 0.1	a 0.00, b 1.0
			♂	0.6 ± 0.1	

[a] Micromoles NADH produced per minute per gram of *Drosophila*.
[b] ♂ and ♀ results combined. n.d., not determined. *a* allele has a high-activity ADH phenotype; *b* allele has a "null" or low-activity ADH phenotype. Based on data for 64 ♂ and 64 ♀ flies.
[c] This species is not discussed in the text but is closely related to *D. inornata*.
Source: Modified from Holmes, Moxon, and Parsons (1980).

utilize ethanol at low concentrations up to a threshold where it becomes toxic for *D. simulans* but remains a resource for *D. melanogaster*; this has been shown experimentally (Table 3.3). A third sympatric cosmopolitan species of the Melbourne area, *D. immigrans*, was found to have an even lower threshold (Table 3.3). These laboratory results relate directly to the field ecology of the three species, since in a pile of grape residues assayed during different decomposition stages immediately following vintage, McKenzie and McKechnie (1979) found only *D. melanogaster* larvae during the fermentation stage when the ethanol concentration was about 7%. At the postfermentation stage when the ethanol concentration was less than 3%, larvae of the sibling species were present. It can, therefore, be inferred that the sibling species can use ethanol as a resource in the wild at low concentrations, and only under high ethanol levels do the two species necessarily occupy different niches (McKenzie and Parsons, 1972).

Fruit and vegetable market results show that the sibling species are fruit specialists, whereas *D. immigrans* utilizes both vegetables and fruits (Atkinson and Shorrocks, 1977). Trapping was carried out in a mixed orchard near Melbourne, Victoria, containing apples, lemons, pears, quinces, peaches, and plums (Prince and Parsons, 1980). Trapping with fermented-fruit baits showed that the proportion of *D. immigrans* adults was significantly higher at lemon tree sites throughout the year (summed data in Table 3.4). A comparison of aspirator catches direct from rotten fruit between lemon and nonlemon sites produced a similar result with an even higher *D. immigrans* proportion. The difference in adult distribution is reflected in the relative larval composition of the fruit, since the proportion of *D. immigrans* larvae from lemons paralleled the aspirator catches.

Lemons, together with other citrus fruits, are unusual in that their decay is rather predictable. Decay is usually associated with infections of the fungi *Penicillium digitatum* or *P. italicum*, which are two fungal species rarely found in nature away from citrus fruits. *D. immigrans* is especially associated with infected fruit, whereas *D. melanogaster* is more often found in uninfected fruit (Atkinson, 1981), and this was confirmed by laboratory experiments with *Penicillium*-infected citrus fruits (Figure 3.6). Clearly, the advantage of *D. immigrans* increases relative to *D. melanogaster*, and also in an absolute sense, as the percent of the surface of each fruit covered with *Penicillium* approaches 100%. The absence of *D. simulans* from Figure 3.6 is explained by the finding that this species emerges from citrus fruits in very low numbers and survives poorly in lemons compared with the other two fly species (Prince and Parsons, 1980; Atkinson, 1981).

Atkinson suggests that *D. immigrans* may have evolved as a citrus specialist in southern China, where the different species of the genus

Table 3.3. *Comparison of threshold concentrations in Drosophila at which ethanol, acetaldehyde, and acetic acid become stresses and values of LT$_{50}$ maximum/LT$_{50}$ control*

Drosophila	Ethanol		Acetic acid		Acetaldehyde	
	Threshold (%)	$\dfrac{\text{LT}_{50}\text{ maximum}}{\text{LT}_{50}\text{ control}}$	Threshold (%)	$\dfrac{\text{LT}_{50}\text{ maximum}}{\text{LT}_{50}\text{ control}}$	Threshold (%)	$\dfrac{\text{LT}_{50}\text{ maximum}}{\text{LT}_{50}\text{ control}}$
melanogaster	11.8	4.1	11.9	2.6	1.2	2.1
Adh^{n2} mutant	1.3	1.6	9.2	3.5	1.0	1.7
simulans	3.4	2.1	5.8	1.6	1.1	2.1
immigrans	1.9	2.4	5.3	1.4	0.5	2.1

Source: Summarized from Parsons and Spence (1981a,b) and Parsons (1982b).

Table 3.4. *Total numbers of* Drosophila immigrans *and non-immigrans flies recorded from lemons and nonlemon fruits, 1973–74*

	Lemon			Nonlemons		
	D. immigrans	Others	Proportion	D. immigrans	Others	Proportion
Traps	2213	710	0.76	2210	5630	0.28
Aspirated	768	46	0.94	860	1091	0.44
Larvae	777	64	0.92	652	1290	0.34

Source: Data from Prince and Parsons (1980).

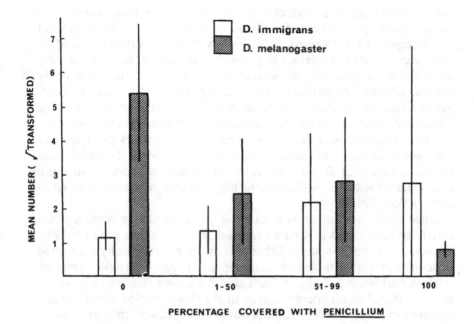

Figure 3.6. The relationship between the percentage of the surface of each fruit covered with *Penicillium* and the mean number of *Drosophila* emerging per fruit. The means are based on sample sizes of at least 19. The numbers were transformed to $(x + 0.5)^{\frac{1}{2}}$ for calculation of 95% confidence limits, and the transformed means are displayed in the figure. (Source: Atkinson, 1981.)

Citrus are native. *D. immigrans* could well have evolved in this area, since most of its close relatives have been reported from the Oriental region (Patterson and Stone, 1952). It may then have spread around the world as these fruits were exploited commercially. It must be stressed, however, that *D. immigrans* breeds on a wide spectrum of fruit and vegetable resources. This may be consistent with its reasonably frequent occurrence at extremely low population densities in floristically diverse rain forests (Brncic, 1970; Parsons, 1981a), where alcohol resource utilization may be less important than in more urban habitats. The comparative study of resource utilization by closely related *Drosophila* species, for example, the comparison of *D. immigrans* with its close relatives, is an open field not only for these species but also for those utilizing more exotic resources.

The cosmopolitan species, *D. melanogaster*, appears to be rather exceptional for its ability to utilize high ethanol concentrations as a resource, although the highest ethanol tolerance of all has been reported in European winery populations of *D. lebanonensis* (David et al., 1979). However, populations of *D. melanogaster* from tropical habitats are less tolerant of high concentrations of ethanol and show lesser larval ethanol preferences than those from temperate habitats (David and Bocquet, 1975; Parsons, 1980a). This shows that the spread of *D. melanogaster* into temperate regions may be associated with natural selection for a premium on ethanol resource utilization, presumably associated with a reduction in the diversity of other resources utilized. Summarizing a number of experiments, *Drosophila* species utilizing fermented fruits in nature may normally have ethanol resource utilization thresholds lower than *D. melanogaster*, especially those species (and populations) from tropical regions. Furthermore, the highly successful invasion of *D. melanogaster* into temperate-zone civilization is associated with the utilization of high ethanol concentrations (Parsons, 1980a, 1981b).

Since the insect cuticle is a selectively permeable membrane penetrated by acetic acid as well as ethanol, gaseous acetic acid could be an additional energy source. Ethanol is in any case normally converted to acetic acid via acetaldehyde and thence to products providing energy (Deltombe-Lietaert et al., 1979). Table 3.3 shows that gaseous acetic acid is utilized as an energy source in the three species under consideration in a parallel way to ethanol utilization. Because they are closely associated metabolically, the concentrations of the two metabolites would be expected to be correlated in nature, so that parallel utilization patterns would be predicted to occur through natural selection. Even so, *D. immigrans* utilizes a much higher concentration of acetic acid relative to *D. melanogaster* when compared with ethanol. This may be consistent with resources apart from ethanol being more important for *D. immigrans* than *D. melanogaster*. In the field, it is likely that in small cavities of fermenting fruits and in other *Drosophila* resources, such as wineries and the rot pockets of cacti, concentrations of ethanol, acetic acid, and other metabolic vapors could reach quite high levels. This means that the capacity of *Drosophila* adults to use nutritive vapors could be very important in nature and additional such volatile compounds, for example, other short-chain alcohols, are likely to be found.

This type of analysis can be taken to the genetic level by comparing the vapor utilization of an *Adh*-null mutant of *D. melanogaster* with the Melbourne population. The biochemical phenotype of the *Adh*-null mutant suggests that adults should have a low ethanol threshold but should utilize acetic acid to levels comparable to that of a wild population. Over ethanol, an Adh^{n2} mutant showed a relatively small longevity increase, which was maximal for 0.5% ethanol (Parsons and Spence, 1981a) with a threshold close to that of *D. immigrans* (Table 3.3). For acetic acid, the Adh^{n2} mutant contrasted dramatically with the ethanol result, in that longevity was greater than the controls over a wide range of concentrations, with a threshold nearly as high as that obtained for the *D. melanogaster* population.

Finally, considering the intermediate metabolic product, acetaldehyde, which is often regarded as highly toxic, low concentrations presumably occur in nature. Accordingly, it is not surprising to find that it is utilized as a resource at low concentrations only (Table 3.3). In this case, interspecific differences were not great, although *D. immigrans* is the most sensitive. The sibling species and the Adh^{n2} mutant are tightly clustered. In any case, differences would be extremely difficult to detect given the toxicity of acetaldehyde at all except the low concentrations here examined (Parsons and Spence, 1981b). Investigations of acetaldehyde levels in *Drosophila* habitats in nature appear necessary.

In addition to longevity under the different ethanol and acetic acid regimes, observations of the reproductive status of flies and the subsequent development of progeny were made. In Table 3.5, results are expressed in terms of the most advanced developmental stage reached (Parsons and Spence, 1981a). Considering ethanol, the *D. melanogaster* population was the fittest since adults emerged over 0.5% ethanol, with development to second-instar larvae up to 9% ethanol. At all except very low concentrations, therefore, ethanol is demonstrably a developmental resource, since there is more development of progeny than for the control. These developmental stages may be achieved by a combination of the use of ethanol and dead flies as resources; in any case such a combined process is likely under field conditions. There is, therefore, a reasonably direct correspondence between this measure of developmental fitness and adult longevity, so that the observation in *D. simulans* of development only to larval stages using up to 3% ethanol is in accord with this pattern. The Adh^{n2} mutant showed development of progeny to the first or second-instar stage for 0.5% ethanol only, which is compatible with Table 3.3.

For acetic acid, the *D. melanogaster* population was again the fittest. In this case, adults developed over 3% and 6% acetic acid, which are somewhat higher concentrations for maximum development than for ethanol. The progeny of the *D. simulans* population did not develop beyond the larval stage as found for ethanol, but development occurred over a wider range as would be predicted from its higher acetic acid threshold. The Adh^{n2} mutant progeny developed to the adult stage over acetic acid, although the developmental stages reached were slightly less advanced than the *D. melanogaster* population at comparable concentrations, which agrees with the somewhat lower theshold of this mutant for acetic acid resource utilization compared with the *D. melanogaster* population.

Finally, the lack of development of *D. immigrans* over both ethanol

Table 3.5. *Developmental stages reached in the offspring of adult Drosophila tested for longevity*

Ethanol/acetic acid concentration (%)	D. melanogaster		D. simulans	D. immigrans
	Wild	Adh^{n2}		
Control	Eggs hatch	Eggs hatch	Eggs hatch	Eggs hatch
Ethanol				
0.01	Eggs hatch	Eggs hatch	Eggs hatch	Eggs hatch
0.03	Eggs hatch	Eggs hatch	Eggs hatch	Eggs hatch
0.1	Pupae	Eggs hatch	Eggs hatch	Eggs hatch
0.5	Adults	1st/2nd instar	Eggs hatch	Eggs hatch
1.5	Prepupa	Eggs hatch	3rd instar	Eggs hatch
3.0	2nd/3rd instar	Eggs hatch	Eggs hatch	Eggs hatch
6.0	Early 2nd instar	—	—	—
9.0	Early 2nd instar	—	—	—
Acetic acid				
0.03	Eggs hatch	Eggs hatch	Eggs hatch	Eggs hatch
0.1	Eggs hatch	Eggs hatch	Eggs hatch	Eggs hatch
0.5	1st/2nd instar	Eggs hatch	Eggs hatch	Eggs hatch
1.0	1st/2nd instar	Eggs hatch	Eggs hatch	Eggs hatch
1.5	2nd instar	2nd instar	1st/2nd instar	Eggs hatch
3.0	Adults	Adult (1 fly)	2nd instar	Eggs hatch
6.0	Adults	Adults (2 flies)	2nd instar	—
9.0	1st/2nd instar	Eggs hatch	Eggs hatch	—

Source: Adapted from Parsons and Spence (1981a).

and acetic acid agrees with the interspecific longevity rankings. Therefore, observations on reproductive status agree well with the longevity data and confirm that longevity of flies exposed to these metabolic vapors is a good indicator of overall fitness.

3.5 Biochemical phenotypes and colonizing potential

The low ethanol threshold of the Adh^{n2} mutant parallels the Australian endemic species *D. inornata* and *D. hibisci*. Therefore, a rare mutant in *D. melanogaster* characterizes perhaps many species exploiting an array of nonfermenting resources. Given that ethanol can be oxidized via catalase and a microsomal oxidizing system (Deltombe-Lietaert et al., 1979), detailed investigations of putatively null mutants and species would repay attention, especially where low thresholds occur as in the Adh^{n2} mutant. Throckmorton (1975) reviewed the phylogeny, ecology, and geography of *Drosophila* and argued that initially the genus was associated with slowly fermenting leaves and other fleshy plant parts on the forest floor, as well as sap, and broken and damaged parts of living plants themselves. This is a relatively austere existence, since such resources are not rich in carbohydrates. However, it may well have provided a step toward the exploitation of the fermentation mode of existence. Therefore, "low" *Adh* activity alleles and low ADH activity (Table 3.2) could be ancestral, and their study could be important for our understanding of noncolonist species.

Considerations of this type were extended to 19 species by Moxon, Holmes, and Parsons (1982). Eleven species of the three subgenera *Drosophila*, *Sophophora*, and *Scaptodrosophila* (see Table 1.2) that are regularly attracted to fermented-fruit baits have higher ADH activities than nine species of subgenera *Hirtodrosophila* and *Scaptodrosophila* that are not attracted to fermented fruit baits (Figure 3.7). In the latter group, the *Hirtodrosophila* species exploit forest fungi; considering the *Scaptodrosophila* species, *D. hibisci* exploits *Hibiscus* flowers, *D. notha* the bracken fern *Pteridium aquilinum*, and the *inornata* complex species are obtained by sweeping in temperate-zone tree–fern forests and sedge/bracken habitats. The subdivision of the *Scaptodrosophila* species into the two ADH activity groups is particularly clear, with the potential colonists in the *lativittata* complex (part of the *coracina* group) all in the higher ADH activity group.

Although there is a substantial published literature on ADH, the enzyme catalyzing the first step of ethanol metabolism in *D. melanogaster*, far less is known concerning the second step. The product of the ADH catalyzed reaction, acetaldehyde, is toxic except at low concentrations (Table 3.3) and must be rapidly transformed into the metabolically usable, nontoxic compound, acetate. A likely candidate for

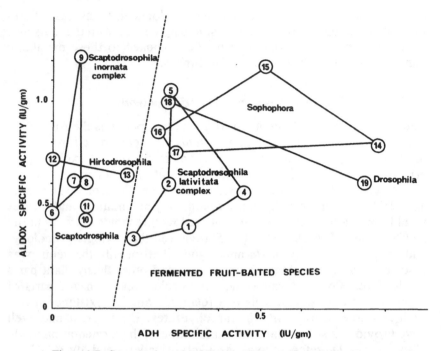

Figure 3.7. Comparative ALDOX and ADH-specific activities among endemic and cosmopolitan Australian species of *Drosophila*. Each point represents an average of results of replicate assays from three separate extracts of 20 adult (10 males, 10 females) flies of each species. *Drosophila* species: Subgenus *Scaptodrosophila*: (1) *lativittata*; (2) *enigma*; (3) *howensis*; (4) *nitidithorax*; (5) *specensis*; (6) *inornata*; (7) *collessi*; (8) *rhabdote*; (9) *fuscithorax*; (10) *hibisci*; (11) *notha*. Subgenus *Hirtodrosophila*: (12) *mycetophaga*; (13) *polypori*. Subgenus *Sophophora*: (14) *melanogaster*; (15) *simulans*; (16) *pseudotakahashii*; (17) *dispar*. Subgenus *Drosophila*: (18) *immigrans*; (19) *hydei*. [Source: Adapted from Moxon, Holmes, and Parsons (1982) where detailed methods are provided for obtaining the extraction buffer.]

catalyzing this reaction is aldehyde oxidase (ALDOX). Although a first expectation may be an association between ADH and ALDOX activities among species, Figure 3.7 shows that the species cannot be subdivided for ALDOX activities as for ADH, so that there is no apparent association of biochemical phenotype with resources for ALDOX. Furthermore, no correlations were observed between ALDOX activity or electrophoretic phenotype and the extent of acetaldehyde resource utilization (Figure 3.8) between the species. Since ALDOX may have a role in *Drosophila* in the detoxification of an array of plant heterocyclic compounds (Moxon, Holmes, and Parsons, 1982) these results may be reasonable. In addition, there is population genetic and electrophoretic

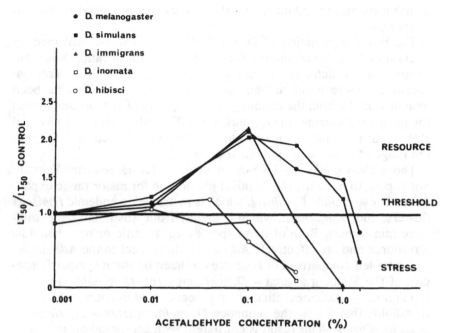

Figure 3.8. Longevity curves for five *Drosophila* species exposed to acetaldehyde vapor expressed as $LT_{50}s$ divided by the $LT_{50}s$ for controls. Thus, when LT_{50}/LT_{50} control >1, acetaldehyde is utilized as a resource and when <1 is a stress. (Source: Adapted from Moxon, Holmes, and Parsons, 1982.)

evidence for at least two structural genes involving ALDOX in a number of Australian *Drosophila* species. Finnerty and Johnson (1979) examined the molecular basis of the extensive electrophoretic variability exhibited by ALDOX in *D. melanogaster*. They concluded that some of the heritable variation in electrophoretic properties of this enzyme is due to other loci that alter ALDOX by processes of post-translational modification. Consequently, variations of these epigenetic processes superimposed upon allelic variation at the structural locus may explain ALDOX levels in natural populations of *D. melanogaster*.

The ALDOX results merely confirm the potential significance of ADH activity as an important biochemical indicator of colonizing ability in *Drosophila*, in combination with electrophoretic phenotypes at the *Adh* locus and ethanol resource utilization.

3.6 Summary

The phenotypes measured by the physical conditions tolerated and the resources utilized by organisms play a vital role in determining their distribution and abundance. For short, they can be called *ecological phenotypes*.

The broad distribution of *Drosophila* species can be explained by tolerances to extreme environmental stresses (desiccation and cold temperature) which impose hard selection upon populations. Tropical species are more sensitive than temperate-zone species. This has been demonstrated within the *melanogaster* subgroup of species and across the genus. Colonizing species must be sufficiently tolerant of physical stresses in new habitats for resource utilization to occur and fall into the range of most cosmopolitan species.

Those *Drosophila* species not attracted to fermented-fruit baits do not appear to have the biochemical phenotype for major range expansions. These include *D. hibisci*, which is restricted to endemic *Hibiscus* flowers, and *D. inornata*, which is mainly from Australian temperate-zone rain forests. Both of these species utilize little or no ethanol as a resource and are effectively *Adh*-null with respect to the *Adh* locus.

A detailed comparison of resources utilized by three sympatric species of the Melbourne area – *D. melanogaster*, *D. simulans*, and *D. immigrans* – indicates ethanol and acetic acid resource-utilization thresholds that follow the sequence *D. melanogaster* > *D. simulans* > *D. immigrans*. This result is consistent with ecological observations on the three species, whereby the former two species are fruit specialists, whereas *D. immigrans* exploits fruits and vegetables. An *Adh*-null mutant of *D. melanogaster* has an ethanol utilization threshold similar to *D. hibisci* and *D. inornata*. The study of such mutants may be important for our understanding of noncolonist species.

The potential significance of ADH activity as an important indicator of colonizing ability at the biochemical level in combination with phenotypes at the *Adh* locus and ethanol resource utilization is confirmed by a comparative study of ADH and ALDOX activities among 19 *Drosophila* species.

4

Variability in natural populations

> Natural selection which guides evolutionary change acts primarily
> on phenotypes, and only secondarily on genotypes. [Waddington,
> 1965:1]

In the last chapter, variation was mainly considered at the interspecific
level. Ecological phenotypes important in determining the distribution
and abundance of organisms were demonstrated. For genetic analysis,
phenotypic variation must be considered at the intraspecific level. Accordingly, in this chapter, variability in natural populations will be
discussed, together with comments on the analysis of such variability.

4.1 Morphology and climate-related variation

Geographical clines exist for morphological characters in widespread
Drosophila species. A thorough study has been conducted on *D. robusta*, a mainly woodland species, which occurs in eastern North
America, extending from southern Canada in the north to the Gulf of
Mexico in the south and as far west as Nebraska. Stalker and Carson
(1947, 1948, 1949) have studied geographical, altitudinal, and seasonal
variation in body size. The geographical study compared 45 strains
collected from 22 widely separated localities for the morphological
measurements of thorax length, head width, femur length, wing width,
and wing length. Significant differences were found between strains
derived from the same locality, but the major variation in the data is
clearly among populations (Figure 4.1). There is a strong correlation
between average annual temperature and body size estimated from a
compound measurement including all five measurements using the discriminant function approach of Fisher (1936). Accordingly, there is a
well-developed geographical north–south cline in morphology, with
body size increasing as the average annual temperature decreases.

Stalker and Carson (1948) then used the discriminant function approach to study an altitudinal transect near Gatlinburg, Tennessee,
from 305 to 1200 m (1000 to 4000 ft) in altitude covering a distance
of 33.3 km (18 miles). Populations from the upper levels of the transect

43

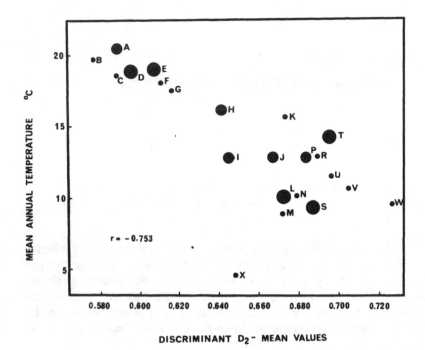

Figure 4.1. Locality mean discriminant (D_2) in *Drosophila robusta* plotted against average annual temperature of locality. Small circles represent localities with only one strain, medium circles localities with two or three strains, large circles represent localities with four or five strains. Letters refer to locality key. A, Austin, Tex.; B, Thomasville, Ga.; C, Marshall, Tex.; D, Abbeville, Ala.; E, Leary, Ga.; F, Verbena, Ala.; G, University, Ala.; H, Morrillton, Ark.; I, Cape May, N.J.; J, Gatlinburg, Tenn.; K, Clinton, Ky.; L, Wooster, Ohio; M, Rochester, N.Y.; N, New Wilmington, Pa.; P, St. Louis, Mo.; R, St. Clair, Mo.; S, Mt. Vernon, Iowa; T, Montauk, Mo.; U, Philadelphia, Pa.; V, Cold Spring Harbor, N.Y.; W, Glenellyn, Ill.; X, Big Fish Lake, Minn. (Source: Stalker and Carson, 1947.)

were significantly more "northern" in their morphology than those from the lower altitudes. Body size increased with increasing altitude, the largest difference occurring between 426 and 610 m (1400 to 2000 ft). Stalker and Carson (1949) studied seasonal variation with a population from St. Louis, Missouri, and detected a regular significant shift toward a "southern" morphology during the summer months. The five morphological measurements, therefore, vary in a quite predictable way, showing that geographical, altitudinal, and seasonal variation of size is associated, directly or indirectly, with temperature.

Similar studies in other widespread species have revealed morphological clines. Body size in the widespread European species *D. subobscura* increases with decreasing average temperature both within

Great Britain and over a larger transect between Scotland and Israel (Prevosti, 1955; Misra and Reeve, 1964). Misra and Reeve (1964) measured their populations after several laboratory generations, whereas Prevosti (1955) showed that body size responded to selection and concluded that the interpopulation differences are largely genetically based. In *D. melanogaster* and *D. simulans*, body weight and ovariole number increase with increasing latitude north of the equator (David and Bocquet, 1975), whereas *D. melanogaster* body size is greater in the cooler uplands than in the lowlands of Puerto Rico (Levins, 1969). It is also of interest that David, Bouquet, and Pla (1976) found that Japanese strains of *D. melanogaster* are more heterogeneous for body weight than French strains of flies, but no interpretation is possible without detailed climatic data. The data appear more difficult to interpret in *D. pseudoobscura*. In one study, Sokoloff (1965) recorded geographical variation in body size not correlated with latitude or altitude, and in a second study (Sokoloff, 1966) yearly variation (related to climatic variation) was found to be greater than geographical variation. Anderson (1968) has reported that natural populations vary in body size from Canada to Mexico such that individuals from Pacific coast populations are smaller than those from inland areas.

From many of these studies, a general correspondence between environment and form emerges, suggesting that selection is the primary cause of ordering of most variation within species over long and short distances (Gould and Johnston, 1972). Taken to its extreme, it suggests that local differentiation is due to the direct effects of climatic selection, even when the organisms concerned are highly mobile (Ehrlich and Raven, 1969). However, the selective advantage of variations in body size among localities remains unclear.

The direct study of altitudinal and latitudinal gradients is worthy of more attention in those parts of the world where there are large climatic variations occurring over short distances within the limits of *Drosophila* resource limitation. It is worth reinforcing the value of an approach used by Clarke (1975) and Borowsky (1977), whereby the role of natural selection can be determined by looking at patterns of variation in two related species for possible geographical correlations in allele frequencies, morphologies, or responses to environmental stresses. The finding that ovariole number and body weight increase with increasing latitude in both *D. melanogaster* and *D. simulans* (David and Bocquet, 1975) suggests similar reactions to temperature-associated environmental variables. However, a cline of increasing adult ethanol resistance with latitude in *D. melanogaster* but only marginally in *D. simulans* is presumably connected with the apparent ability of the former species to utilize relatively high concentrations of ethanol in temperate regions (Section 3.4), and so the effects of natural selection appear as a major cline in one of the species only (Stanley and Parsons, 1981).

4.2 Climatic stresses

The discussions on morphological variation agree with the conclusion (Section 3.2) that the success of a *Drosophila* population depends basically upon its adaptation to annual climatic cycles – a view favored by Andrewartha and Birch (1954) from considerations of insect populations in general, especially those in temperate regions. For both of the extreme stress categories of high-temperature/desiccation and low temperature, it is important to distinguish between conditions for resource utilization, which involves feeding and breeding, and survival, although they would be expected to be correlated to some extent (Section 7.2).

Considering survival (or mortality) rates, high-temperature-sensitive strains of *D. melanogaster* from the wild are known, and heterogeneity among isofemale strains (strains set up from single inseminated females from the wild) has been described for ability to withstand desiccation and high-temperature shocks (Parsons, 1980a). Populations of both *D. melanogaster* and *D. simulans* from Uganda, an area of high temperatures, are more resistant to high temperatures than are population samples from less extreme environments (Tantawy and Mallah, 1961). Temperature "races" occur in *D. funebris* from northern Europe, northern Africa, and Asiatic Russia, such that resistance to high and low temperatures corresponds to climate (Dubinin and Tiniakov, 1947). These cosmopolitan species, therefore, provide evidence of being subdivided into "races" defined with respect to extreme environments.

Mortality after exposure to the cold stress of $-1°C$ has been measured for isofemale strains of *D. melanogaster* and *D. simulans* from Townsville, northern Queensland (a subtropical climate with a hot, damp summer and warm, dry winter, at latitude 20°S, Figure 4.2), and Melbourne (a temperate climate with a hot, relatively dry summer and cool, damp winter, at latitude 38°S, Figure 4.2). There was significant variation among isofemale strains within populations (Parsons, 1977a). Although there was some overlap among populations, the sequence of mean mortalities was *melanogaster* (Townsville) > *simulans* (Townsville) > *simulans* (Melbourne) ≫ *melanogaster* (Melbourne). In particular, the *simulans* populations did not differ significantly, whereas the *melanogaster* populations did ($P < 0.001$). In Melbourne, resource utilization effectively ceases during winter months, since activity, feeding, and mating of these species effectively ceases at about 12°C and below (McKenzie, 1975), whereas there is no such restriction in the

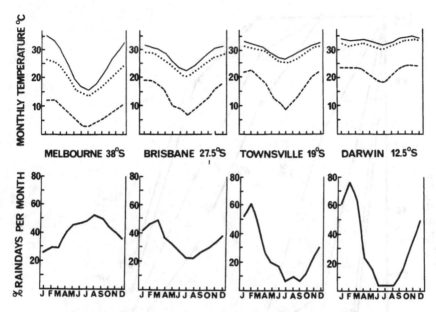

Figure 4.2. Mean monthly temperature and rainfall profiles for four Australian study sites. ····, mean maximum temperature; ——, 86th percentile of maximum temperature (exceeded only 1 day in 7), ---, 14th percentile of minimum temperature (lower only 1 day in 7). (Source: Modified from Stanley and Parsons, 1981.)

Townsville population as can be seen from the meteorological data in Figure 4.2.

Overall, the population rankings for cold-resistance suggest similar responses for the two species, implying that similar selective forces are acting upon them. In addition, the climatic data in Figure 4.2 suggest that the Melbourne populations may be more tolerant of high-temperature/desiccation stress than Townsville. This is simply because Melbourne has a hot, relatively dry summer, compared with Townsville, where the summer is damp with less extreme heat. Laboratory tests for high-temperature stresses lasting 6 hr at 0% and 95% humidity confirm the prediction (Parsons, 1980c). Figure 4.3 shows that Townsville mortalities exceed Melbourne mortalities for four paired contrasts, the intraspecific differences being smaller than the interspecific differences. Parallel selective forces, therefore, occur for tolerance to both low- and high-temperature stresses in both species. Note that these physiological tests are in accord with Figure 3.2, where the clear advantage of *D. melanogaster* over *D. simulans* when temperature fluctuations are highly variable is demonstrated. The advantage of *D. me-*

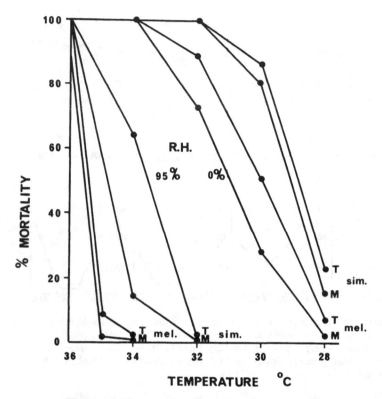

Figure 4.3. Percent mortalities of *Drosophila melanogaster* (mel.) and *D. simulans* (sim.) after 6 hr at various temperatures for 95% RH (left-hand group of four plots) and 0% RH (right-hand group of four plots). M, Melbourne; T, Townsville. (Source: Adapted from Parsons, 1980c.)

lanogaster over *D. simulans* on these tests can almost certainly be correlated with the earlier buildup of *D. melanogaster* in spring in Melbourne compared with *D. simulans*, leading to a maximum population of *D. melanogaster* in late spring and of *D. simulans* in late summer (McKenzie and Parsons, 1974a). Cold stress evidently involves differing underlying physiological (and genetic) processes from desiccation stress, since the correlation coefficient among isofemale strains for mortalities to the two stresses is close to zero. Indeed, levels of resistance to desiccation stress are associated with body weight and metabolic rate, but this is less likely for resistance to cold (Parsons, 1977a). Isofemale strains provide a quick way of assessing the possibility of associations among traits in natural populations, which can then be investigated for underlying causal relationships (see Section 4.4).

Since temperature is among the most important of factors determin-

Table 4.1. *The relative contribution of the three largest chromosomes (X, II, III) of* Drosophila melanogaster *to resistance of cold temperature at each of five developmental stages*

Eggs	II > III > X
Second-instar larvae	III > X, II
Third-instar larvae	III > II > X
Pupae	II > III > X
Adults	III > II > X

Source: After Marinković, Tucić, and Kekić (1980).

ing *Drosophila* species distributions, Marinković, Tucić, and Kekić (1980) investigated whether the same or different genes determine resistance to cold temperature at different life-cycle stages in *D. melanogaster*. In addition, any results may relate to the issue as to which, if any, life-cycle stage is the most important in overwintering. Marinković, Tucić, and Kekić (1980) exposed *D. melanogaster* to low temperatures permitting at most 10% to 15% of individuals to survive, from which subsequent generations were set up. Initial exposure levels were 0°C/14 hr for eggs, 0°C/26 hr for second-instar larvae, 0°C/30 hr for third-instar larvae, − 3°C/14 hr for pupae, and − 3°C/28 hr for adults. Responses to selection occurred in all cases, and the stress levels were increased after the fifteenth, twenty-fifth, and thirty-ninth generations. After 52 generations of selection, that is, after almost 4 years, cold-resistance levels had increased 123 times above the initial value in eggs, 75 times in second-instar larvae, 73 times in third-instar larvae, 81 times in pupae, and 558 times in adults. Corresponding to these figures are realized heritabilities (measures of the proportion of variability of a trait that is under genetic control – see Appendix) of 0.067 for eggs, 0.047 for second-instar larvae, 0.041 for third-instar larvae, 0.073 for pupae, and 0.14 for adults (Tucić, 1979). Second chromosome genes were mainly involved in controlling the cold resistance of eggs and pupae, and third chromosome genes, for larvae and adults (Table 4.1), whereas the X chromosome was of lesser importance. These results indicate that this particular population of *D. melanogaster* has an enormous capacity for adaptation to low temperatures, especially the adult stage. Comparisons of *D. melanogaster* with closely related noncolonists, as in Section 3.2, are likely to be important for our understanding of the colonization potential of the cosmopolitan sibling species *D. melanogaster* and *D. simulans*. In any case, a high level of flexibility for adaptation is suggested by the differing genes involved at different life-cycle stages.

4.3 Benign and stressful habitats

Detailed analyses of variability have been carried out in five Australian populations, three from the east coast – Melbourne, Brisbane, and Townsville – and two populations extremely close to each other from the extreme north – Darwin and Melville Island (Figure 1.1). Melbourne has a temperate climate with periods of extreme heat and cold stress, whereas Darwin and Melville Island in the tropics also have periods of extreme heat stress. By comparison, Brisbane and Townsville have humid subtropical and tropical climates, respectively, with lesser heat stresses. Hence, in terms of heat stress, the most widely separated populations are subject to greater levels of stress than the intervening populations (Figure 4.2).

Assuming the importance of climatic selection, higher desiccation/heat resistance would be expected for the extremes, that is, the Melbourne and Darwin/Melville Island populations should show greater desiccation resistance than the Brisbane and Townsville populations. Percentages alive after exposure to 0% RH at 25°C for various time intervals were ranked as follows (Parsons, 1980d):

$$\left.\begin{array}{l} \text{Darwin} \\ \text{Melville I.} \\ \text{Melbourne} \end{array}\right\} > \text{Townsville} > \text{Brisbane}$$

in accordance with expectation.

Development time would be expected to be more rapid when the climate is extreme than under more optimal climates. In confirmation of this prediction, the ranking of development times was

$$\left.\begin{array}{l} \text{Darwin} \\ \text{Melville I.} \\ \text{Melbourne} \end{array}\right\} < \text{Townsville} < \text{Brisbane}$$

where the difference between extremes amounted to about 0.4 day (Stanley and Parsons, 1981). In view of earlier comments on development time (Chapter 2), these differences are probably consistent with the climatic contrasts among the four populations.

Each of these five populations was made up of 8 to 10 strains set up from single inseminated females or isofemale strains, permitting the assessment of intrapopulation variation (Parsons, 1980d). Because of strong selection for high-desiccation resistance and rapid development times in the extreme populations, they would be expected to be less variable than populations from more optimal habitats. Broadly speak-

Table 4.2. *Isofemale heritabilities of mortality after desiccation and development time for five Australian populations of* Drosophila melanogaster

	Latitude (°S)	Climate	Mortality after desiccation		Percent emergence time	
			Males	Females	Day 11	Day 12
Darwin ⎱ Melville I. ⎰	12	Hot, dry winter; hot, humid summer	−0.16 0.09	0.15 0.22	0.40 0.11	0.02 0.19
Townsville Brisbane ⎰	19 ⎱ 27 ⎰	Warm, dry winter; warm, humid summer	0.63[b] 0.70[b]	0.58[b] 0.72[b]	0.29[a] 0.54[b]	0.44[a] 0.63[b]
Melbourne	37	Cool, wet winter; hot, relatively dry summer	0.11[a]	−0.18	0.52[b]	0.27

Note:
Variation among isofemale strains significant at [a] $P < 0.01$,   [b] $P < 0.001$.
Source: Calculated from Parsons (1980d).

ing, this was found to be true, especially for desiccation resistance, implying strong directional selection for these traits. Correspondingly, *isofemale heritabilities* (proportion of variability among isofemale strains that is under genetic control – see Appendix) tend to be highest in the more optimal populations (Table 4.2). This is very clear for desiccation resistance, since in populations from stressful extreme habitats isofemale heritabilities are low, whereas in populations from benign habitats they are extremely high. For development time, isofemale heritabilities are of similar magnitude in all populations except for Darwin and Melville Island, where variation among isofemale strains was insignificant (Parsons, 1980d), and isofemale heritabilities tend to be lower than in the remaining populations. The Melbourne climate, however, is almost certainly such that there would be less intense selection for rapid development time than in Darwin and Melville Island. In the latter localities, the extreme and continuous high-temperature stress throughout the year would mean that resources would normally be short-lived, imposing continuous selection in the one direction.

As will be seen in Section 5.3, quite high heritabilities are likely for environmental stress traits, such as desiccation. However, the lack of variability in the extreme latitude populations suggests that capacity for adaptation to even more extreme habitats may be restricted. It is in benign habitats where variability is high since extremes are not continuously favored. The occasional high isofemale heritabilities of development time provide evidence that this trait may in certain circumstances be quite variable in natural populations (see also Chapter 2).

The number of sympatric *Drosophila* species commonly attracted to fermented fruit baits in four of the localities was: Townsville, 6; Brisbane, 4; Melbourne, 3; and Darwin, 2. Assuming interspecific competition to be unimportant, as is likely at the population level in species so near the *r* end of the *r*–*K* continuum (Wiens, 1977), high species diversity presumably corresponds to heterogeneous available resources as preliminary ecological surveys indicate (Stanley and Parsons, 1981). Based upon the universal use of ethanol as a resource by those *Drosophila* species attracted to fermented fruit baits (Parsons, 1981a), it can be predicted that ethanol-utilization levels and their variability would be a reflection of their habitats in nature. It has already been noted that the ethanol-utilization threshold is greater in temperate than in tropical zone populations of *D. melanogaster* (Section 3.4). In addition, even though average thresholds of *D. simulans* and *D. immigrans* are much lower, they show the same trend across climates, so that ethanol tolerance presumably reflects a general characteristic of *Drosophila* resources, which on average may contain less ethanol in tropical compared with temperate zones (Stanley and Parsons, 1981).

For the east-coast populations, variability among isofemale strains for ethanol tolerance is in the sequence Townsville > Brisbane > Melbourne, so that tolerances fall and become more variable toward the equator (Table 4.3). The Darwin population has a somewhat higher ethanol tolerance than is predicted from its latitude although that of Melville Island is lower. These are ecologically marginal populations with high-desiccation resistances and rapid development times, where ethanol tolerance can be regarded as a direct reflection of the limited available resources. Resource homogeneity would be expected as indicated by low variability among isofemale strains. (Table 4.3).

Another indication of resource utilization comes from reactions of newly hatched larvae to ethanol based upon the argument that this is when maximum feeding occurs in nature. In these experiments, 10 newly hatched larvae are placed centrally on agar in a petri dish (Figure 4.4). One semicircle of the agar contains one of the ethanol concentrations, the other is pure agar. A range of concentrations was selected, whereby the threshold between attraction and avoidance (see Dethier, Barton-Browne, and Smith 1960; Shorey, 1977) after 15 min could be identified at high concentrations, and the approximate point of no response at low concentrations. Using 6% ethanol, which is between these extremes, the east-coast preferences are in the same sequence as the adult ethanol tolerances, with Darwin and Melville Island having the lowest preferences in this case (Table 4.3). The isofemale variability and heritability rankings are similar to those for ethanol tolerance with

Table 4.3. *Adult ethanol tolerance and larval preference for ethanol for five Australian populations of Drosophila melanogaster*

	No. of isofemale strains tested	LT_{50} for adults on 12% ethanol (hr)	F values for variability among isofemale strains	Isofemale heritability	No. of larvae out of 10 choosing 6% ethanol	F values for variability among isofemale strains	Isofemale heritability
Darwin	10	41	8.2[b]	0.59	5.4	2.2	0.13
Melville I.	10	19	6.7[b]	0.53			
Townsville	9	9	70.8[b]	0.93	6.4	4.7[b]	0.32
Brisbane	9	29	12.0[b]	0.69	6.8	3.1[a]	0.21
Melbourne	8	60	8.3[b]	0.59	8.2	1.6	0.07

Note:

Adult tolerances are expressed as mean LT_{50}s exposed to 12% ethanol in hours, based upon five replicates of 25 flies per sex per isofemale strain. Larval preferences are given as means of the number of newly hatched larvae out of 10 choosing agar containing 6% ethanol given a choice of plain agar and ethanol-containing agar (eight replicates per isofemale strain).

[a] $P < 0.01$. [b] $P < 0.001$.

Source: Calculated from data in Parsons (1980d,e).

Figure 4.4. Procedure for testing responses of *Drosophila* larvae to metabolites. Ten newly hatched larvae are placed in the center of the petri dish on the boundary between agar and agar plus metabolite, in this case 6% ethanol. After various periods of time, usually 15 min, the numbers of larvae on the two sides of the petri dish are scored and preferences calculated.

the Brisbane and especially the Townsville populations being highly variable among isofemale strains (Parsons, 1980d,e). Although the ranking of isofemale heritabilities for the two traits is similar, overall values are lower for the behavioral trait compared with the resource utilization trait, showing that the latter trait is the better one for detailed genetic analyses.

Based upon the $r-K$ continuum model, an overall summary (Table 4.4) shows that ecological marginality means a shift toward the r end of the $r-K$ continuum associated with resource-utilization homogeneity. Therefore, life history, environmental stress, and resource-utilization traits, which are phenotypic assessments directly relatable to the field, provide some understanding of ecologically marginal populations. The probability of a widespread species spreading further may be assessed by an investigation of such phenotypes especially in species not restricted by resource limitations at the margins. If a population is phenotypically extreme associated with low variability, then it is likely that an ecological margin has been approached from which colonization of more extreme habitats is unlikely. Indeed, such a population can be regarded as analogous to one at a plateau following many generations of artificial directional selection, although rare recombinant events may occasionally lead to greater extremes usually with lower fitness

Table 4.4. *A comparison of the characteristics of the five Australian* Drosophila *populations considered in Tables 4.2 and 4.3*

Characteristics	Melbourne, Melville I., Darwin	Brisbane and especially Townsville
Desiccation resistance	High with low variability	Low with high variability
Development time	Fast with low variability	Slow with high variability
Ethanol tolerance	Relatively homogeneous	Relatively heterogeneous
Larval preference for ethanol	Relatively homogeneous	Relatively heterogeneous
Overall summary	Tending toward r selection	Tending toward K selection

(Mather, 1943; Thoday and Boam, 1961). Field evidence for lowered fitness in marginal environments comes from mosquito fish that in a small fresh-water stream had fewer large young and made smaller reproductive efforts, than fish living 150 to 300 m away in a more optimal, large brackish estuary (Stearns and Sage, 1980). Even so, there is always a possibility of spreading into even more extreme habitats provided that resources are available. However, Milkman's (1978) limited success in selecting for increased adult heat resistance in *D. melanogaster* suggests that large increases in tolerance to extremes may on occasions be difficult to achieve. In agreement are experiments on *Dacus tryoni*, where Meats (1981) failed to produce changes in thresholds for cold torpor or cold survival, after 2 years of selection. However, Marinković, Tucić, and Kekić (1980) successfully selected for cold resistance in *D. melanogaster* as just mentioned, and in the hymenopteran, *Aphytis lingnanensis*, laboratory selection for survival of temperature extremes produced marked changes in temperature tolerance (White, DeBach, and Garber, 1970). Ultimately, however, the basic physiology of the organism will become restrictive, and in some cases it appears that populations are quite close to this limit.

4.4 Isofemale strain analyses

Populations made up of isofemale strains were used for comparisons at the geographical level in the aforementioned study. In addition, by the use of simple procedures, such as diallel crosses (a set of all possible matings between strains), isofemale strains can provide an indication of comparative genetic architectures among populations. Furthermore, comparisons among species can be made as shown by studies of desiccation resistance in populations of *Drosophila melanogaster* and its sibling species *D. simulans* from Brisbane and Melbourne based upon diallel crosses (McKenzie and Parsons, 1974b). When desiccation re-

Table 4.5. *Summary of diallel analyses of resistance to desiccation for Melbourne and Brisbane populations of* Drosophila melanogaster *and* D. simulans *indicating the number of significant components (P < 0.001) out of eight for the Melbourne populations and four for the Brisbane population*

Source of variation[a]	Melbourne		Brisbane	
	melanogaster	*simulans*	*melanogaster*	*simulans*
a	8	8	4	4
b	8	0	4	4
c	0	0	0	0
d	0	0	0	0
Maximum entry possible	8	8	4	4

[a] a, Tests primarily additive effects; b, tests dominance effects; c, tests average maternal effect; d, tests the remainder of the reciprocal variation.
Source: Simplified from McKenzie and Parsons (1974b)

sistance was studied on an annual basis, only the Melbourne *D. simulans* population showed cyclical changes in mean mortality such that the population was most resistant in summer, becoming less so as the weather becomes cooler. Only in this population was additive genetic variation solely significant (Table 4.5), so that seasonal adaptation is presumably due to gene frequency changes following directional selection in one direction in one season (e.g., for resistance to high temperatures) and in the other season in the other direction (e.g., for resistance to cold temperatures). Such oscillating directional selection could well simulate stabilizing selection and would be manifested by a high level of additive genetic control (Parsons, 1973a). In the other populations where dominance/epistasis was additionally significant, a greater proportion of the genome is apparently involved with interaction components. An understanding of genotypes determining the distribution and abundance of *Drosophila* species in this way should enable predictions of possibilities of colonization events. Isofemale strains are valuable in this regard, since information is obtainable irrespective of our knowledge of the genetics of a species, provided that it can be cultured in the laboratory.

Isofemale strains, therefore, permit the assessment of phenotypic variation at the *population* level (Parsons, 1980a). The power of the approach in the study of phenotypic variation derives from the persistence over many generations of consistent strain differences originating from the founder females. The wild-inseminated females can be regarded genetically as "pairs," although it is now known that nearly half the *D. melanogaster* females derived from natural populations may carry the sperm of more than one male (Milkman and Zeitler, 1974), and a similar situation occurs in *D. pseudoobscura* (Anderson, 1974). However, this does not affect the approach in a qualitative sense, whereby isofemale strains can be regarded as families under test within a population.

Refinements now permit the claim that the *isofemale-strain approach* constitutes one of three approaches to quantitative inheritance. The *biometrical approach* had its origins in Fisher's (1918) paper, where it was shown that the genetics of biometrical or quantitative traits parallels that of Mendelian traits. This approach has been presented in its modern form by Kempthorne (1957), Falconer (1960), and Mather and Jinks (1971). Independent of this biometrical approach, Payne (1918) used directional selection to locate genetic activity for scutellar chaeta (bristle) number in *D. melanogaster* to the chromosome level. Beginning mainly with Thoday (1961), many genes (polygenes) controlling quantitative morphological traits, such as bristle number and wing vein length, have been located in the last 20 years. A major conclusion is that the genetic architecture of many phenotypic traits – especially morphological traits and environmental stresses – comprise a rather low number of structural genes with relatively large, mainly additive effects. The novel feature of this *polygene-location approach* is that it integrates the use of chromosome markers to trace sections of selected chromosome subdivided by recombination with the method of progeny testing as used in the analysis of quantitative traits. Although Thoday's work was based on directional selection lines, isofemale strains can also be used to localize genetic activity to the chromosome (and gene) level (Parsons, 1980a).

In fact, the *isofemale-strain approach* merely takes some of the simpler features of the biometrical and polygene-location approaches, in order to study the range and nature of phenotypic variation in natural populations (Parsons, 1980a). Although it was initially developed with morphological traits, it is of particular value for ecobehavioral traits as shown in the previous section.

In the context of this book then, the isofemale approach is useful because:

1. Rapid and quick inferences can be made on quantitative traits and their genetic architectures for populations of any species that can be cultured in the laboratory from environments ranging from optimal to extreme and for traits ranging from molecular and physiological to ecological and behavioral.

2. Responses to directional selection may be accelerated by basing selection on extreme isofemale strains, a result of impor-

Table 4.6. *Correlation coefficients between various parameters* (df *= 16 in all cases) for female data based upon 18 isofemale strains from a natural population of* Drosophila melanogaster *collected in Victoria, Australia*

	Mean weights after desiccation, B	B/A	Mean dry weights	Numbers dead (angularly transformed)
Mean wet weights, A	0.784[a]	0.597[b]	0.791[a]	−0.579[c]
Mean weights after desiccation, B		0.961[a]	0.705[b]	−0.902[a]
B/A			0.578[c]	−0.926[a]
Mean dry weights				−0.436

[a] $P < 0.001$.   [b] $P < 0.01$.   [c] $P < 0.05$.
Source: Simplified from Parsons (1970).

tance for the location of polygenic activity and in an applied context.

3. Correlation studies for pairs of traits based upon many isofemale strains may provide information on the possibility of shared developmental or physiological processes versus linkage between polygenes. Comparisons can be made of life-cycle stages, for example, larvae and adults, and of extreme stresses (including toxic chemicals). For example, Parsons (1970) found substantial differences among 18 isofemale strains collected from Leslie Manor, Victoria, Australia, for desiccation stress measured by mortalities after 16 hr in a dry environment, indicating large genetic differences in this population; those strains with high wet and dry weights lost water relatively less rapidly and had lower mortalities than those with lower wet and dry weights as indicated by the correlation coefficients in Table 4.6. In the field, Levins (1969) found that those flies taken at midday were 5% larger than those collected in the morning or late afternoon, which suggests that only those flies that are genetically most resistant to desiccation by virtue of their large size appear at midday. It is difficult to imagine that the correlations in Table 4.6 are due wholly to linkage between segregating polygenes because of their magnitude.

4. Since isofemale strains provide assays of phenotypic variation within species, comparisons among species can be accurately made, for example, in the comparative study of the ecobehavioral genetics of *D. melanogaster* and *D. simulans* (Parsons,

1975a). Some of the parameters that can be computed with the isofemale approach are discussed at the end of the Appendix. They mainly involve computing components of variance at the population level and related parameters so as to obtain estimates of the isofemale heritability (Tables 4.2 and 4.3).

It may be thought that the inbreeding levels involved in setting up isofemale strains may be a problem. Certainly, genetic variability of a population is expected to decline when it goes through a bottleneck. However, the reduction in average heterozygosity per locus depends not only on the size of the bottleneck, but also upon r, the Malthusian parameter (Nei, Maruyama, and Chakraborty, 1975). Indeed, if the population size increases rapidly after going through a bottleneck, the reduction in average heterozygosity is rather small even if the bottleneck size is extremely small. Thus, for $r = 0.5$ and a bottleneck where $N = 2$ (as in an isofemale strain without multiple insemination), the reduction in heterozygosity is about 50%. This still means that an isofemale strain has basically a genetic architecture of an outbred population, especially in colonizing high-r species. In fact, this is likely to be an absolute maximum reduction in heterozygosity for two reasons: (1) multiple insemination so that N is effectively >2, and (2) the likelihood of heterozygote advantage especially in the stressful environments likely to be encountered by colonizing populations (Section 5.4). In other words, isofemale strains certainly do not have the highly homozygous genetic architectures of the inbred strains that form the basis of some models of quantitative inheritance (e.g., Mather and Jinks, 1971). The small variability fall appears consistent with each isofemale strain exhibiting its own "identity," which has been shown to persist for many generations (Parsons, 1980a). It is this feature that makes isofemale strains an extremely useful tool in the analysis of phenotypic variation at the population level. This depends on the aforementioned conclusion that the traits under analysis are mainly under the control of a relatively low number of structural and largely additive genes, which will be further discussed (Sections 5.3 and 6.3). Indeed little divergence among isofemale strains would be expected under a genetic architecture of many polygenes all of small effect – a commonly assumed model in quantitative inheritance.

In *D. melanogaster*, an isofemale strain is a laboratory population that has undergone a single bottleneck, where $N = 2$ as a minimum, followed by population sizes of the order of $N = 50$ to 100. It is important for genetic analyses at the population level that isofemale strains retain their variability levels for sufficient generations to permit analyses that can be related back to the population in the wild. Following the simplified approach in Frankel and Soulé (1981), an approximation of the amount or proportion of variance that remains following the sudden reduction of a large population to a small one containing N individuals is

$$1 - \frac{1}{2N}$$

and after t generations the expected proportion of genetic variance remaining is

$$\left(1 - \frac{1}{2N}\right)^t$$

However, in an isofemale strain N is not constant over generations, since $N = 2$ for the bottleneck generation and then becomes larger.

The effective size of a population where the number per generation varies with time is the harmonic mean of the effective number of each generation, or,

$$\frac{1}{N_e} = \frac{1}{t}\left(\frac{1}{N_1} + \frac{1}{N_2} + \cdots + \frac{1}{N_t}\right)$$

After the bottleneck generation where $N = 2$, 75% of the genetic variance remains, that is, 25% of the variance is lost. After 10 generations, for $N = 10$ in subsequent generations to the bottleneck, 52% of the variance remains, for $N = 50$ the figure becomes 71% and for $N = 100, 74\%$. Hence, for population sizes in excess of 50, little genetic variability is lost after the bottleneck. In addition, because of the likelihood of multiple insemination and heterozygote advantage, these are likely to be minimum estimates. Accordingly, an isofemale strain should reflect the initial founder inseminated female for a substantial period of time. For example, for both scutellar bristle number and resistance to high-temperature shocks, isofemale strains have been shown to retain their characteristics for in excess of 40 generations (Parsons and Hosgood, 1967; Hosgood and Parsons, 1968). At the molecular level, the stability of isofemale strains has been recently demonstrated by Coen, Thoday, and Dover (1982). They found that the genetic behavior of rDNA structural variants in separate and mixed populations derived from isofemale strains is stable for over 1000 generations in *D. melanogaster*. This suggests that the variants are not subject to strong selection under the relatively optimal conditions of the laboratory.

In addition, observations on scutellar bristle number for flies kept at four temperatures ranging from 15° to 27.5°C for 20 to 50 generations show that the founder female effect is preserved across environments (Hosgood and Parsons, 1971). Table 4.7 gives percentages of males

Table 4.7. *Percentages of* Drosophila melanogaster *with more than four scutellar bristles in three isofemale strains at four temperatures*

Temperature (°C)	Strain	Generation						
		5	10	15	20	25	30	35
15[a]	1	22	4	7	17			
	2	6	3	—	—			
	3	0	0	0	0			
20	1	10	12	15	10	20	12	19
	2	4	8	6	2	3	1	3
	3	0	0	0	1	0	0	0
25	1	3	4	7	3	6	8	3
	2	1	4	3	0	1	1	1
	3	0	0	0	0	0	0	0
27.5	1	1.4	1.0	1.9	0.4	0.6	1.0	2.3
	2	1.2	1.0	0.3	0.4	0.3	0	0
	3	0	0	0	0	0	0	0

[a] The low number of generations at this temperature is because of long generation times. Strain 2 died out before generation 15.
Source: Adapted from Hosgood and Parsons (1971).

having more than four scutellar bristles for three isofemale strains. In all cases, the ranking is $1 > 2 > 3$ ranging from strain 1 with many flies with more than four bristles especially at the lower temperatures, to strain 3 where bristle number is rigidly canalized to four bristles. The higher bristle numbers at low temperatures is consistent with other published data (e.g., Rendel and Sheldon, 1960; Pennycuik and Fraser, 1964). However, Table 4.7 shows that this environmental effect does not swamp the isofemale strain effect. It would be particularly interesting to carry out similar experiments on traits having definite ecological importance, such as resistance to high temperature and desiccation, which may be expected to be under intense selection pressure in the occupation of new habitats.

It should finally be noted that Nei, Maruyama, and Chakraborty (1975) demonstrated that the average number of alleles per locus is profoundly affected by bottleneck size. This is because random genetic drift eliminates many low-frequency alleles, a phenomenon of little consequence for the analysis of quantitative traits among isofemale strains, especially if carried out soon after their introduction into the laboratory. The procedure usually adopted in the *Drosophila* laboratory at La Trobe University of testing isofemale strains as soon as possible, and very rarely beyond 10 generations, appears therefore to be quite safe, since most genetic variability will be retained although some rare alleles may be lost.

4.5 Summary

Climate has a major effect upon morphology, in particular, body size. As one extreme, local differentiation may be due to the direct effects of climatic selection, even when the organisms concerned are highly mobile.

The success of a *Drosophila* population depends basically upon its adaptation to annual climatic cycles. Cosmopolitan species provide evidence of being subdivided into "races" defined with respect to the extreme environments occurring during climatic cycles. In addition to resistance to environmental stresses, other traits including development time and utilization of ethanol as a resource can be related to the climate from which populations are derived.

Comparisons among populations are greatly assisted by studying iso-female strains (derived from single-inseminated founder females) from natural populations and the variability among them. The isofemale-strain approach is illustrated by an attempt to relate the adaptations of populations of *D. melanogaster* derived from different climates to the predictions of the $r-K$ continuum model. In summary, the isofemale-strain approach takes some of the simpler features of the biometrical and polygene-location approaches to quantitative inheritance in order to study the range and nature of phenotypic variation in natural populations. Although there is a bottleneck at the stage when isofemale strains are set up, the subsequent reduction of variability in a species such as *D. melanogaster* is trivial, so that the isofemale strains directly reflect the common genes of the founder inseminated females for many generations.

5

Genetic variability, ecological phenotypes, and stressful environments

The likelihood of organismic, instead of merely genic, selection goes far toward meeting one of the most serious objections to the theory of natural selection encountered by Darwin. [Wright, 1980:841]

5.1 Central and marginal populations

In the last 25 years, a substantial literature on genetic variation in central and marginal populations based upon gene (in particular, electrophoretic) and chromosome polymorphisms has been built up. Much of this information is based upon central and marginal habitats defined in a geographical sense, in contrast with the usually correlated ecological dichotomy of benign versus stressful.

Considering electrophoretic variants, levels of polymorphism generally do not vary much between central and marginal populations in several widespread *Drosophila* species. These include *D. pseudoobscura* (Prakash, Lewontin, and Hubby, 1969), *D. robusta* (Prakash, 1973a), *D. willistoni* (Ayala, Powell, and Dobzhansky, 1971), and *D. subobscura* (Marinković, Ayala, and Andjelković, 1978). Lewontin (1974) argued that genic heterozygosity is high in marginal populations, because the temporal instability of such environments means that no particular genotype is favored for long periods. This is supported by laboratory experiments in *D. willistoni* and *D. pseudoobscura*, indicating that heterozygosity is higher in heterogeneous than in constant environments (Powell, 1971; Powell and Wistrand, 1978). In addition, certain natural population surveys are suggestive of slightly higher gene variability levels in marginal populations, for example, in *D. robusta* (Prakash, 1973a) and *D. melanogaster* (Band, 1975); this agrees with the temporal (and, hence, ecological) instability interpretation. Extrapolation is, however, needed for other organisms, as suggested by Nevo and Yang's (1979) finding of greater genetic diversity in central populations of tree frogs in Israel. Here, climatic factors, primarily water availability, are shown to be the best predictors of polymorphism and heterozygosity levels. Another example is the iguanid lizard, *An-*

olis grahami, in which there are more polymorphic loci and slightly higher individual heterozygosity in the parental Jamaican population compared with colonists in Bermuda (Taylor and Gorman, 1975). The climate of Bermuda is more seasonal than Jamaica so that climatic selection is a possibility, especially as not all the alleles examined are selectively neutral.

In one of the earliest discussions, Townsend (1952) found that marginal populations of *D. willistoni* have somewhat lower frequencies of autosomes producing recessive adverse effects on viability and fertility as homozygotes than do central populations. In these experiments, chromosomes were made homozygous, so it is not possible to determine whether a single gene is responsible or several in combination. The equilibrium frequencies of lethals and other deleterious chromosomes are theoretically lower in small than in large populations, so that somewhat lower frequencies may be expected in marginal than in central populations of a species. This is certainly a feature of populations of *D. melanogaster* undergoing serious annual shrinkages in Hungary compared with neotropical populations in Colombia (Hoenigsberg et al., 1969). A relatively low proportion of lethals and semilethals also tends to occur in ecologically marginal populations of *D. pseudoobscura* (Dobzhansky et al., 1963).

Band (1963) compared the frequency of lethals and semilethals in *D. melanogaster* from South Amherst, Massachusetts, in (1) a constant 25°C, (2) a fluctuating temperature with a wide range of 14°C between the daily maximum and daily minimum, which approximates the wider range of temperatures observed in the natural environment, and (3) a fluctuating temperature with an 8°C range. The frequency of lethals and semilethals was highest in (3), so that both the extremes of a constant and wide-ranging temperature regime lead to a reduction of genetic diversity. Indeed, coupled with Beardmore (1961), who found a higher additive genetic variance in *D. melanogaster* for sternite bristle number at 25° ± 5°C than at 25°C, and data on the fitness of heterozygotes for the same comparison, it can be concluded that genetic diversity is highest in favorable environments, where fluctuations are insufficiently extreme to be stressful.

In later experiments, Band and Ives (1968) found that lethal and semilethal frequencies were significantly correlated with total summer rainfall over the period 1938–64, that is, when the climate is drier, the frequencies are lower. This result is plausible on the assumption that the drier environment is more marginal for *Drosophila*, which follows from the known sensitivity of *Drosophila* to desiccation stress (Sections 3.2 and 4.3).

Turning now to chromosome inversion polymorphisms, in certain species notably *D. willistoni* (da Cunha, Burla, and Dobzhansky, 1950), *D. robusta* (Carson, 1955; Prakash, 1973a), *D. funebris* (Dubinin and Tiniakov, 1947), and *D. subobscura* (Lakovaara et al., 1976), marginal populations tend to be less variable than central ones. Considering *D. robusta* in more detail (Carson, 1958), populations may be characterized by means of an index of free recombination, taken as the average percent of the genome available for crossing-over in females from nature. A consistent relationship between this index of free recombination and the location of populations was found (Figure 5.1), whereby the index increased from low values in central populations to higher values at the margins. Central populations of *D. robusta* tend to have indices of about 70%, intermediate populations about 78%, and marginal populations 80% to 100%. This cytological variation is not related to obvious environmental variables, and Carson (1965) suggests that low inversion heterozygosity may be favored in populations of ecologically marginal habitats, because it permits greater flexibility via genetic recombination. Experimental data in support include Carson's (1958) finding that monomorphic populations of *D. robusta* responded more quickly to selection for "motility" than did polymorphic ones and Markow's (1975) finding in *D. melanogaster* that the presence of inversions on chromosomes 1, 2, and 3 reduced the effectiveness of selection for negative phototaxis compared to strains with fewer inversions. The experimental results of Tabachnick and Powell (1977), whereby chromosomally monomorphic populations of *D. willistoni* adapt to chemical stresses better than polymorphic populations, are also consistent with Carson's interpretation.

Arguing mainly from *D. willistoni*, Dobzhansky and his colleagues (da Cunha and Dobzhansky, 1954; da Cunha, Burla, and Dobzhansky, 1950) suggested that the gene arrangements of inversions have an ecotypic function, whereby at the center of the distribution of a species, diverse habitats are occupied so favoring a diversity of gene arrangements compared with ecological margins, that is, there may be a mosaic of karyotypes varying in incidence and frequency according to microhabitat in ecologically central populations. Under adverse conditions, near climatic limits where there may be a limited flora with few fruits, inversion frequencies tend to be low (da Cunha, Burla, and Dobzhansky, 1950) compared with nonstressful more central regions with a varied fruit-bearing flora offering more ecological niches. This rationale is not inconsistent with Carson's interpretation. A third interpretation (mentioned earlier) stresses the temporal instability of marginal environments (Lewontin, 1974). Variation in selection at the margins and during colonization phases must be almost certainly greater than at the center. This interpretation permits the selection of a number of distinct

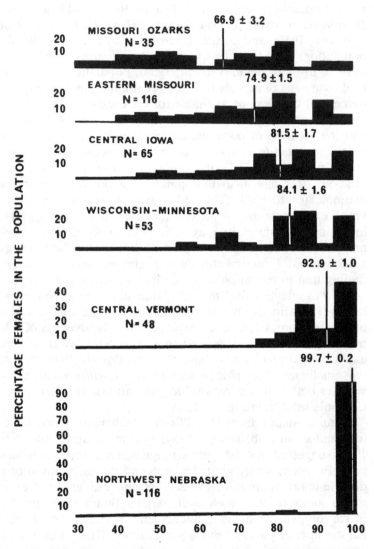

Figure 5.1. Correlation of distribution and free recombination in natural populations of *Drosophila robusta*. The index of free recombination (1) in the Missouri Ozarks is considered typical of central populations, (2) in Eastern Missouri, Central Iowa, and Wisconsin–Minnesota, typical of intermediate populations, and (3) in Central Vermont and especially northwestern Nebraska, typical of marginal populations. (Source: Carson, 1958.)

Table 5.1. *Summary of conclusions on comparative variability levels in central and marginal populations of certain* Drosophila *species*

	Central	Marginal
Genes (especially electrophoretic variants)	Equivalent	Equivalent, but occasionally higher
Lethals/semilethals (genes and chromosomes)	Higher	Lower
Chromosomal polymorphism	Higher	Lower

Note:
The text provides an indication of exceptions.

and diverse modes in central populations and is compatible with the necessity for recombination in marginal habitats and so incorporates the first two interpretations.

The aforementioned observations show that in some *Drosophila* species electrophoretic polymorphism levels are as high at the margins as in central populations, whereas inversion polymorphism levels are lower at the margins (Table 5.1). An inversion is a coadapted complex consisting of both structural and regulatory genes. In ecologically marginal habitats, directional selection may favor new or rare combinations determined by a minority of available inversions, so that such populations will become depleted of inversions and epistatic combinations generally (Soulé, 1973). The remaining structural-gene alleles at polymorphic loci may then permit direct responses to extreme environments to occur. This may not directly involve electrophoretic loci but could involve linked loci, so that electrophoretic variants may be affected as a "hitch-hiking" effect (Maynard Smith and Haigh, 1974; Thomson, 1977), but no overall electrophoretic variability change would be expected. Assuming that the number of genes for resistance to an extreme stress, for example, desiccation, is relatively few as appears likely (Section 4.4), it would be advantageous that these genes should be readily combinable together in populations in marginal habitats to enable adaptation to more extreme conditions in as few generations as possible. Hence, inversion polymorphisms that may lock such loci together may not be favored in such habitats. Similarly, the buildup of epistatic combinations leading to chromosomes that are lethal or semilethal as homozygotes would be less likely under such conditions (Table 5.1).

Even so, marginal populations are prone to severe reductions in numbers and so in theory gene diversity might be expected to fall as found in marginal populations of vertebrates, for example, tree frogs, iguanid lizards, and other species cited in Soulé (1973). However, just a trickle of gene flow can restore lost alleles. This may explain why marginal

Table 5.2. *Proportion of polymorphic loci out of 43 examined, average number of alleles per locus, and proportion of loci estimated to be heterozygous in an average individual for* Drosophila pseudoobscura *populations*

Population	Proportion of polymorphic loci	Average number of alleles per locus	Proportion of loci heterozygous
Strawberry Canyon, Calif.	0.535	2.116	0.163
Mesa Verde, Colo.	0.488	2.000	0.125
Austin, Tex.	0.465	1.907	0.133
Bogotá, Colombia	0.256	1.349	0.068
Mean excluding Bogotá	0.496	2.007	0.140

Source: Simplified from Prakash (1977a).

Drosophila populations retain most of their allelic diversity. One extremely marginal population of *D. pseudoobscura* from the Andes of Colombia does, however, show low variation for inversion polymorphisms, lethals, and semilethals (Dobzhansky et al., 1963), and electrophoretic polymorphisms (Prakash, 1977a) (Table 5.2); in fact, this population is sufficiently divergent that Ayala and Dobzhansky (1974) regard it as a separate subspecies. The low variation presumably resides in founder effects and inbreeding associated with extremely low population sizes (Nei, Maruyama, and Chakraborty, 1975) and possibly periodic population crashes. Bryant, van Dijk, and van Delden (1981) returned to this problem when discussing the introduction of the face fly, *Musca autumnalis*, into the United States from Europe. They found that the average heterozygosity and the average number of alleles per locus was less, but not significantly so, in the American flies than in ancestral European flies. Their paper provides a useful discussion of problems of interpretation when the population sizes at the bottleneck stage are unknown. More experiments along the lines of a transfer (Holdren and Ehrlich, 1981) of a montane Wyoming population of the checkerspot butterfly, *Euphydryas gillettii*, to the Colorado mountains are needed, especially where population sizes and the genetic composition of the colonists are known.

Implicit in this discussion is that populations in marginal habitats are subjected to quite severe climatic selection and range expansions involve increased tolerance to extreme environmental stresses. This is also likely in the land snails, *Cepaea nemoralis* and *C. hortensis*, since at the northern and southern limits and other margins (e.g., altitudinal), the incidence of color morphs appears to depend more upon climatic

selection (temperature, shading, humidity) at the microhabitat level
rather than upon habitat diversity (Jones, 1973a,b; Bantock and Noble,
1973; Bantock and Price, 1975; Arnason and Grant, 1976). At the interspecific level, three species of the land snail, *C. nemoralis, C. hortensis*, and *Arianta arbustorum*, differ in survival, rate of water loss,
and behavior under conditions of low humidity in laboratory studies
(Cameron, 1970) – a result that could have intraspecific parallels. Bantock and Price (1975), studying marginal populations of *C. nemoralis*,
argue that minimum night temperature is the limiting factor. Indeed,
under temperature extremes in open vegetation, the degree of polymorphism is depressed compared with woodland populations. They
argue that under suboptimal environments small climatic variations
may be critically important. It must be stressed, however, that climate
is but one of eight forces postulated to affect shell polymorphism in
Cepaea and ''almost nowhere will it be possible to explain the observed
pattern of genetic variation by involving only one of them'' (Jones,
Leith, and Rawlings, 1977). A recent comparative study reveals little
association of pattern of geographical variation at the visible and molecular levels (Jones, Selander, and Schnell, 1980). Assuming variation
in shell polymorphism patterns is largely due to natural selection, then
such selection must be weak or absent at the electrophoretic polymorphism level, that is, assessments at the visible phenotypic level
may be of primary importance in understanding the evolution of local
races.

5.2 Ecological phenotypes

The results in land snails imply that in order to understand the genetic
structure of natural populations, it is important to consider the organism as the basic unit of selection, and not just genic and chromosomal
selection (see Wright, 1980). In this book so far, the organism has been
looked at from the ecological point of view. In parallel, Johnson (1979a)
has developed the concept of a physiological phenotype developed
especially from considerations of plant adaptation.

In all likelihood, a concern of greater fundamental significance than
variability levels among genomes is the nature and role of those loci
directly affecting ecologically important phenotypic variables that are
easily assayable, so that genetic analyses can be readily carried out.
Physical stresses have been discussed in some detail. Ethanol tolerance
will be considered briefly here, because of considerable discussions in
the literature for possible associations with the alcohol dehydrogenase
(*Adh*) locus and alcohol dehydrogenase (ADH) enzyme activity especially in *Drosophila melanogaster*. Indeed, when the level of com-

parison comprises different species classified by resource-utilization differences, there is a fairly direct association of ethanol tolerance, *Adh* genotype, and ADH enzyme activity (Section 3.3).

Ethanol tolerance is an ecological phenotype, since it relates to resources in nature. It can also be regarded as a metabolic phenotype in the sense of Johnson (1979b) which is controlled by several loci presumably including the *Adh* locus. In wild *D. melanogaster* at the Chateau Tahbilk winery in Australia, there is, however, little association between ethanol tolerance and genotypes at the *Adh* locus (McKenzie and Parsons, 1974c; McKenzie and McKechnie, 1978). The relation between this locus and environmental ethanol is, therefore, obscure [see Gibson and Oakeshott (1980) for a recent overview of this complex area], even though there is some laboratory evidence for ethanol-mediated selection at this locus (Gibson, 1970). The single-locus approach cannot in itself explain the maintenance of genetic variability in natural populations, even though the importance of environmental selective agents, including temperature (Pipkin, Rhodes, and Williams, 1973), is occasionally demonstrable at this level. Arguing from human data, temperature effects may be quite common although often difficult to detect (Piazza, Menozzi, and Cavalli-Sforza, 1981); these authors give data suggesting that two-thirds of 35 enzymes or proteins show significant associations with climate. Returning to *Drosophila*, the ecological approach is to begin with ethanol tolerance and assay the contribution of the *Adh* locus and ADH enzyme activity to this phenotype, rather than the more usual approach of commencing with the electrophoretic locus and seeking correlations of adaptive significance. The adaptive significance of ethanol tolerance in nature is emphasized by the observation of parallel clines in the three species *D. melanogaster, D. simulans,* and *D. immigrans,* whereby ethanol utilization and tolerance increase as habitats become more temperate (Section 4.3), and from the interspecific considerations mentioned earlier.

Temperature, temperature-correlated variables perhaps relating to resources, or a combination of the two may be important in the frequently reported *Adh* allele frequency clines from flies collected in nature (Malpica and Vassallo, 1980). In addition, temperature and ethanol concentration have major effects upon the longevity of adults exposed to ethanol vapor (Ziolo and Parsons, 1982), but do not much influence ADH activity (Figure 5.2). The data in Figure 5.2 are based upon the genotypes $Adh^F Adh^F$ (FF), $Adh^S Adh^S$ (SS), and the hybrid $Adh^F Adh^F$ male $\times$ $Adh^S Adh^S$ female (FS) and its reciprocal (SF). Longevity is the measure of ethanol tolerance using the apparatus in Figure 3.4 and is expressed at $LT_{50}s$ (Section 3.3). ADH activity was measured using standard assays similar to those described by Day, Hillier, and Clarke (1974). For ethanol tolerance, the four main effects – genotype,

Table 5.3. *Analyses of variance for ethanol tolerance measured by LT$_{50}$ values and for ADH activity in* Drosophila melanogaster

| | Degrees of freedom | Ethanol tolerance | | ADH activity | |
		Mean square	F	Mean square	F
Temperature (T)	3	128.53	313.71[a]	6.49	8.46[a]
Ethanol (E)	6	51.12	124.78[a]	3.52	4.59[a]
Genotype (G)	3	10.07	24.58[a]	283.73	370.31[a]
Sex (S)	1	28.62	69.86[a]	25.28	32.99[a]
T × E	18	4.41	10.77[a]	1.23	1.60
G × T	9	0.60	1.47	0.89	1.16
G × E	18	0.54	1.33	2.38	3.10[a]
S × T	3	3.60	8.79[a]	0.18	0.23
S × E	6	1.00	2.43	1.47	1.92[a]
S × G	3	0.18	0.44	2.70	3.53
G × T × E	54	0.44	1.07	1.06	1.38
S × T × E	18	0.21	0.51	1.47	1.92
S × G × T	9	0.03	0.08	0.63	0.82
S × G × E	18	0.41	0.99	0.80	1.05
S × G × T × E	54	0.19	0.45	1.09	1.48
Error	1208, 544[b]	0.41		0.77	

[a] $P < 0.001$.
[b] The first number is the degrees of freedom for ethanol tolerance, and the second for ADH activity.
Source: Data from Ziolo and Parsons (1982).

temperature, ethanol concentration, and sex – were significant, in particular, temperature and ethanol concentration (Table 5.3). The four main effects were significant for ADH activity, but in contrast with ethanol tolerance, the genotype effect was by far the largest as shown in Figure 5.2, where FF flies had over twice as much activity as SS flies with heterozygotes being intermediate. Hence, environmental variables affecting one set of data in a major way do not greatly affect the other set.

There are conflicting results in the literature; on the one hand, FF homozygotes have been shown to be less tolerant of ethanol vapor, as in Figure 5.2, and on the other hand, the opposite may occur where survival in ethanol-supplemented media or exposure to media impregnated with ethanol is used as the test situation (Day, Hillier, and Clarke, 1974; Kamping and van Delden, 1978; Oakeshott et al., 1980; Ziolo and Parsons, 1982). Irrespective of these complications, adult longevity in an ethanol atmosphere does appear to be a trait of ecological significance, which is mainly affected by temperature and ethanol concentration. This laboratory result is consistent with field results from

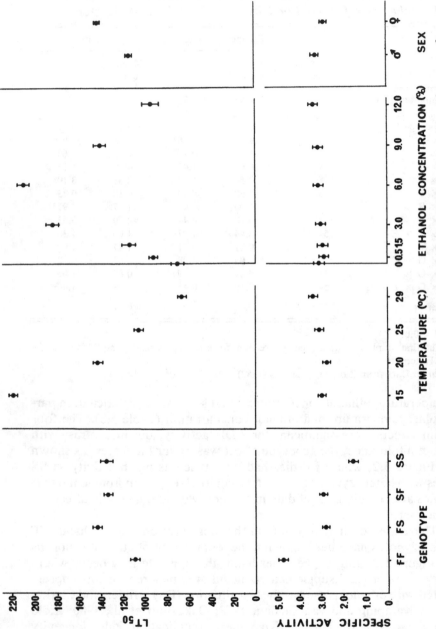

Figure 5.2. Ethanol tolerances measured as LT_{50}s in hours and ADH enzyme activities in *Drosophila melanogaster* for genotypes, temperatures, ethanol concentrations, and sexes. Vertical lines represent 95% confidence limits. (Source: Ziolo and Parsons, 1982.)

the Chateau Tahbilk winery (McKenzie and Parsons, 1974c; McKenzie and McKechnie, 1978). It emphasizes the need to select phenotypes of physiological and ecological importance as primary study materials, perhaps using interspecific comparisons as a guide to the selection of appropriate phenotypes.

It is, therefore, important to assess how ostensible ecological phenotypes vary within and among populations and the degree to which this variation is under genetic control. In genetically well-known organisms, such as *D. melanogaster*, genetic activity can be localized to chromosomes and regions of chromosomes by applying the polygene-location approach to the study of variability in natural populations (Thoday, 1961; Thompson and Thoday, 1979) to extreme isofemale strains (McKenzie and Parsons, 1974c). At this stage, the real involvement of electrophoretic phenotypes with ecological phenotypes should become apparent in a way that would be difficult to obtain using electrophoretic phenotypes as the primary assessment of populations. Comparisons among populations consisting of several isofemale strains should, therefore, enable the genetic dissection of both ecologically marginal and colonizing populations in a biologically meaningful way. In many situations, ecological phenotypes should form the basic data, which is an approach contrasting with that of commencing with electrophoretic polymorphism data. Even so, Johnson (1979b) gives a number of examples of physiological phenotypes, where the single-locus approach provides some information; however, key elements in understanding the total physiological phenotype at the organismic level may of course be missed. This is simply because any single-locus approach ignores multiple loci, degree of linkage, and any interactions between loci.

Finally, temporal instability has been used as an operational definition of ecological marginality, coupled in the case of Australian populations of *D. melanogaster* with the number of sympatric species. Perhaps this biological approach is more widely applicable in attempting to understand ecological phenotypes, especially if the distributions of closely related species are compared [Wilson (1965) has elaborated a similar approach with ants]. For example, the *obscura* species subgroup is represented in North America by species such as *D. pseudoobscura*, *D. persimilis*, *D. miranda*, and *D. lowei*. Of these, *D. pseudoobscura* has the widest distribution. In regions of geographical overlap of *D. pseudoobscura* and *D. persimilis*, the former species is more frequent in warmer and drier locations and at low elevations (Figure 5.3), whereas the latter is common in opposite circumstances (Dobzhansky and Powell, 1975b). It is not then surprising that *D. pseudoobscura* has been found in artificial habitats, for example, orchards. Information on the genetic composition of *D. pseudoobscura* popula-

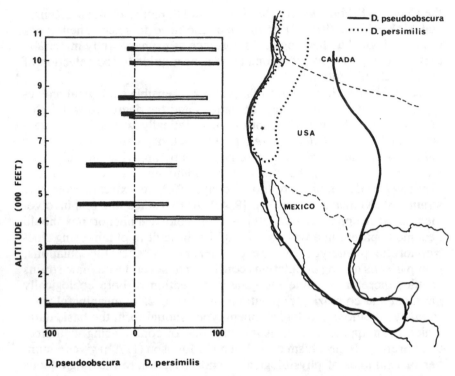

Figure 5.3. The relative proportion of *Drosophila pseudoobscura* and *D. persimilis* at different altitudes along a 70-mile transect in Yosemite Calif. (see asterisk on map), together with the approximate distributions of the two species. (Source: Wallace, 1981.)

tions in disturbed habitats is scarce, but such information as is available on chromosome polymorphisms indicates no differences compared with natural habitats (Dobzhansky, 1965), as is also true of recolonization after fire (Moore, Taylor, and Moore, 1979). Interestingly, compared with earlier observations, the less widespread species, *D. persimilis* and *D. miranda*, appear to be undergoing range expansions (Moore, Taylor, and Moore, 1979). Therefore, although analyses of sympatric species may be useful in defining marginality, a certain fluidity of species distributions may occur. In particular, there is a tendency for a number of geographically widespread species to extend rather than contract their ranges (Parsons and Stanley, 1981). Curiously, there is some evidence in several parts of the world that *D. simulans* is expanding at the expense of the most widespread species of all, *D. melanogaster* (Watanabe and Kawanishi, 1976; Parsons and Stanley, 1981). Analysis of ecological phenotypes at the intra- and interspecific levels are needed for a better understanding of these range changes.

5.3 Stressful environments and genetic analysis

There is a slowly increasing literature devoted to comparing the effects of optimal and extreme environments for traits demonstrably having a role in the determination of fitness. For example, Parsons (1978a) asked whether there is necessarily a direct association between longevity under optimal and stressful environments. In other words, would those individuals surviving longest under an optimal environment do so under a stressful environment, defining the environment to include both biological and sociological factors?

Aging in natural populations has hardly been considered except in humans, where heredity and environment are almost impossible to disentangle. In *Drosophila*, most longevity studies have been carried out on inbred strains, their hybrids, and mutants. One of the few studies on natural populations is that of Parsons (1978a) on the longevity of 20 isofemale strains in *D. melanogaster* and *D. simulans* at each of the temperatures, 20° and 25°C. In *D. melanogaster*, differences among strains were found with a high correlation among strains across temperatures showing longevity to be under genetic control. In *D. simulans* there were differences among strains at each temperature associated with a large strain × temperature interaction such that the correlation among strains across temperatures was effectively zero. From these results, studies on the genetics of aging only provide information with respect to the environment chosen for study. Although the two environments are ostensibly similar for the two species, in fact they are not, since *D. melanogaster* is readily cultured at both temperatures, whereas it is less easy to culture *D. simulans* at 25°C (Section 3.1). This suggests that if the environments are not stressful, there may be some correspondences across environments. It is certainly not surprising that studies on the genetics of aging have not proceeded far, especially as both temperatures are encountered in nature although not, of course, continually. Both species will, in fact, survive in excess of 6 hr at 30°C without loss of fertility provided the humidity is high (Parsons, 1980c, Figure 4.3).

This rather surprising result gains support from data on ^{60}Co-γ-irradiation in *D. melanogaster* showing no association between radiosensitivity to the extreme dose of 110 krad and longevity of untreated flies for six isofemale strains and from similar results for four inbred lines (Westerman and Parsons, 1972). Presumably there are differing genes or gene complexes determining longevity alone compared with

Table 5.4. *Mean longevities (in days) of inbred lines ($\bar{P}$) and hybrids ($\bar{F}$) based upon 4 × 4 diallel crosses in* Drosophila melanogaster *for exposure to ^{60}Co-γ-irradiation between 0 and 120 rads and a hybrid superiority measure expressed as ($\bar{F}_1 - \bar{P}_1$)/($\bar{F}_1 + \bar{P}_1$)*

	Dose (krad)					
	0	40	60	80	100	120
Males						
$\bar{P}_1$	32.86	20.15	14.01	10.97	7.68	2.21
$\bar{F}_1$	40.65	24.76	16.80	13.60	10.35	3.10
Females						
$\bar{P}_1$	29.23	20.99	18.57	14.86	9.65	3.94
$\bar{F}_1$	41.65	35.86	28.03	22.07	15.60	6.75
($\bar{F}_1 - \bar{P}_1$)/($\bar{F}_1 + \bar{P}_1$)						
Males	0.1059	0.1026	0.0906	0.1072	0.1479	0.1676
Females	0.1752	0.2615	0.2030	0.1953	0.2359	0.2627
Significance level						
gca	<0.01				<0.05	<0.001
sca	<0.05	<0.001	<0.01	<0.001	<0.001	<0.001
gca/sca	1.76	0.55	0.65	0.27	0.56	7.32

Note:
Levels of significance for general combining ability (gca) measuring additive effects and specific combining ability (sca) measuring nonadditive effects are given with ratios of the mean squares for gca and sca.
Source: Derived from data of Westerman and Parsons (1973).

longevity after γ-irradiation. The genetic basis of exposure to ^{60}Co-γ-irradiation has been investigated by Westerman and Parsons (1973) at varying irradiation levels based on a diallel cross (the set of all possible matings between several strains) between four inbred lines (Table 5.4). From such crosses, it is possible to calculate the general combining ability (gca), which provides an estimate of the additive genetic variance (see Appendix), and the specific combining ability (sca), which provides an estimate of nonadditive (mainly dominance) variance (Griffing, 1956; Parsons, 1967). Mean longevity decreases with each successive dose-rate increase, so that at 120 krad mean longevity is of the order of one-tenth of the control. At the extreme 120-krad stress, the additive genetic variance is highly significant, and the nonadditive variance while significant at $P < 0.001$ is minor by comparison as shown by the ratio of the mean squares for gca and sca (Table 5.4). At all other doses, the nonadditive variance is more important, whereas the control data show significant additive and nonadditive variances, neither being large. Therefore, the genetic basis of longevity varies according to the level of environmental stress – assessed in this case by longevity after irradiation.

There is no reason why isofemale strains should not give similar results; indeed, a high level of additivity was obtained in the range of 100 –120 krad (Parsons, MacBean, and Lee, 1969). Given this level of additivity, gene localization studies have been shown to be practicable using isofemale strains; such studies have been taken to the chromosome arm level in chromosomes II and III (see Parsons, 1980a) and regions of major additive gene effects detected. The genetic basis of the polygenic trait longevity can, therefore, be most readily analyzed under extreme environmental stress conditions. This conclusion also applies to other stresses including those of a chemical nature (Crow, 1957; Parsons, 1980a). One exception as indicated in Table 4.2 is for desiccation resistance, which in extreme environments shows little variation, presumably because of continuous selection for increased desiccation resistance in nature. However, this is unlikely to be so for most populations, except those from extreme habitats. Unless there are direct correlated effects, novel stresses to which populations have not been previously exposed would be likely to show considerable additivity.

The death rates at these extremely high doses of ^{60}Co-γ-rays parallel those that may be obtained under the hard selection imposed by desiccation or temperature stress. From this, it may follow that colonizing populations may be relatively simple to analyze genetically, because of their tendency to experience severe environmental stresses, even if sporadically.

In Table 3.3, it was shown that *D. melanogaster* utilizes ethanol and acetic acid vapor up to threshold concentrations, where they become stresses. This offers the opportunity to compare variability among isofemale strains in optimal and stressful situations. Accordingly, LT_{50} values were obtained over 3% and 12% ethanol and over 6% and 12% acetic acid. The lower concentration in each case corresponds closely to that at which the metabolites are utilized as resources to a maximal level, whereas the higher values are somewhat above the threshold concentrations in the Melbourne population. Means are given expressed as the ratios LT_{50}/LT_{50} control in Table 5.5. As expected from earlier considerations (Section 4.3), longevities fall in the sequence Melbourne > Brisbane > Townsville for both metabolites as shown by the decreasing LT_{50}/LT_{50} control ratios. In all cases, the higher concentration is stressful, since the LT_{50}/LT_{50} control ratios are <1.

Isofemale heritabilities (Section 4.3; and Appendix) are much greater over the higher concentrations of ethanol and acetic acid than those obtained over the lower concentrations. Indeed, the heritabilities over

Table 5.5. *Mean LT$_{50}$/LT$_{50}$ control values and isofemale heritabilities from analyses of variance within and between nine isofemale strains of* Drosophila melanogaster *for Melbourne, Brisbane, and Townsville populations exposed to ethanol and acetic acid vapor*

	Ethanol		Acetic acid	
	3%	12%	6%	12%
Melbourne				
Mean	3.02	0.96	1.93	0.89
Heritability	0.38[a]	0.68[b]	0.33	0.68[b]
Brisbane				
Mean	2.99	0.52	1.54	0.39
Heritability	0.35[a]	0.56[b]	0.35[a]	0.76[b]
Townsville				
Mean	2.34	0.49	1.46	0.39
Heritability	0.16	0.82[b]	0.46[b]	0.87[b]

Note:
Variation among isofemale strains significant at [a] $P < 0.01$, [b] $P < 0.001$.
Source: Calculated from data of Parsons (1982b).

the higher concentrations are all >0.5 and over the lower concentrations <0.5. In addition, the two highest heritabilities are for the Townsville population over the higher concentrations of the metabolites. This is the population with the lowest LT$_{50}$/LT$_{50}$ control ratios, that is, where the stresses are greatest of all. Hence, the level of variation among isofemale strains depends upon the environment; that is, whether it is stressful or nonstressful. The expressed variation among phenotypes is, therefore, highest under conditions of environmental stress and appears to increase with the level of stress provided that populations have not previously experienced such stresses (Section 4.3). In any analysis of the genetic basis of ethanol and acetic acid differences in natural populations, it would obviously be simplest to begin where the metabolites are stressful.

Derr (1980) studied variation in life-history characteristics of the American cotton stainer bug, *Dysdercus bimaculatus*, from the genetic point of view. This species is a colonist in the tropics living on short-lived seed crops maturing at various times throughout the tropical dry season (Section 2.1). Using parent–offspring data, heritabilities were obtained for life-history traits relating to time of egg production including the timing of the first clutch and the interclutch interval; both had an estimated additive genetic variance, V_A, and hence a heritability close to zero. This implies that directional selection acts to minimize these measures in a colonizing species utilizing unpredictable resources. In contrast, traits relating to the quantity of eggs produced, such as the size of the first clutch, total number of eggs, and the number of clutches, gave a V_A of about 0.30. At first sight, directional selection would also be expected to minimize V_A for these traits. Derr attributed the differences in the proportion of additive genetic influences for the two trait categories to differences in the way timing and quantity affect productivity, and variability among individuals in genotype–environment interactions for clutch size. Derr then investigated tolerance to moisture stress, because the uncertainty of reproducing at a particular place and time in the habitat of this species leads to the prediction that some females would be more stress tolerant than others (as in *Drosophila*). For timing of the first clutch, the additive genetic variance became 0.30 to 0.40, indicating that the population could respond to directional selection at times of environmental stress – a result of probable importance in adaptation to new habitats. In addition, this agrees with the earlier conclusion that genetic analysis is simplest in populations exposed to severe environmental stress conditions – such as may occur during colonization episodes – and demonstrates yet again that the expressed genetic variability depends upon the environmental conditions.

5.4 Genetic heterogeneity and fitness

In experimental animals (and plants), fitness differences between heterozygotes and heterokaryotypes on the one hand, and homozygotes and homokaryotypes on the other, are frequently maximal in extreme environments favoring the heterogeneous genotypes (Figure 5.4). Although Figure 5.4 involves desiccation stress, much of the published data are on the effects of temperature extremes. It has been argued that such extreme-environment heterosis (Parsons, 1971) is a consequence of the greater versatility or efficiency conferred to heterozygotes by the presence of different genes (Schwartz and Laughner, 1969; Fincham, 1972; Johnson, 1974; Singh, Hubby, and Lewontin, 1974). Alternatively, the homozygotes have poorer fitness than the corresponding heterozygotes because of the breakdown of coadaptation in forming homozygotes; this breakdown is magnified in extreme environments. Crossing between inbred lines in several *Drosophila* species typically leads to heterosis, which is magnified under extreme environments (Figure 5.5). Similarly, levels of hybrid superiority increase with irradiation dose especially for the comparison between the control and 120-krad data as in Table 5.4.

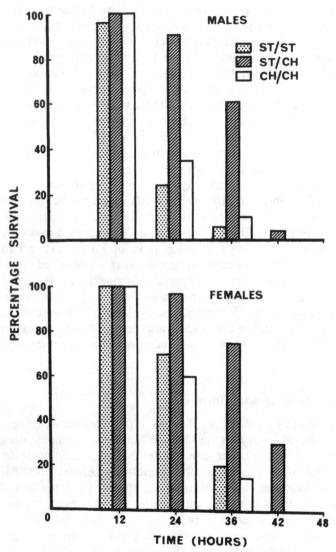

Figure 5.4. Percentage survivals of males and females for the ST/ST, ST/CH, and CH/CH karyotypes of *Drosophila pseudoobscura* when desiccated. (Source: Adapted from Thomson, 1971.)

Of interest at the natural population level is the contrast between individuals of differing heterozygosity levels, rather than the contrast between artificially derived inbred strains and their hybrids. The earlier evidence does, however, suggest that individuals with high heterozygosity levels should be fitter than those with lower levels (although the possibility of associations between environmental and genetic heter-

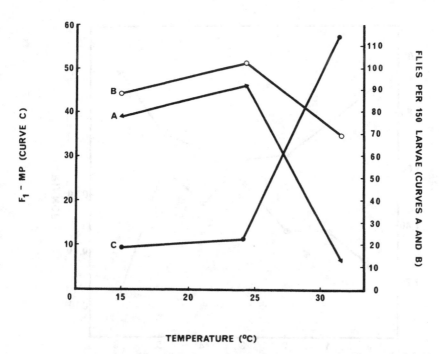

Figure 5.5. Heterosis at three temperatures for larval survival in *Drosophila melanogaster*. A = mean of midparent, MP; B = mean of all F_1s; C = B − A. (Source: Parsons, 1973b.)

ogeneity must not be ignored). At the natural population level, comparisons of highly heterozygous individuals with those that are less so, for traits directly or indirectly contributing to fitness, should provide a test of this hypothesis (see also Frankel and Soulé, 1981).

In the American oyster, *Crassostrea virginica*, Zouros, Singh, and Miles (1980) took a cohort, which they weighed a year later, and examined the oysters electrophoretically at seven loci. Dividing the sample into weight classes, they found that an individual's weight is positively correlated with the number of loci for which it is heterozygous (Figure 5.6). Additionally, the variance in weight is lower among individuals of high degrees of heterozygosity. Overdominance in growth rate appeared to Zouros, Singh, and Miles to be the most plausible explanation for the correlation between weight and degree of heterozygosity. To what extent growth rate relates to overall fitness is an open question, although larger individuals may enter reproduction sooner and produce more gametes, so suggesting such a relationship.

A generalization of this type of result to plants comes from Schaal and Levin (1976) from work on the perennial self-incompatible herb, *Liatris cylindracea*. Plants were divided into six age classes, and the

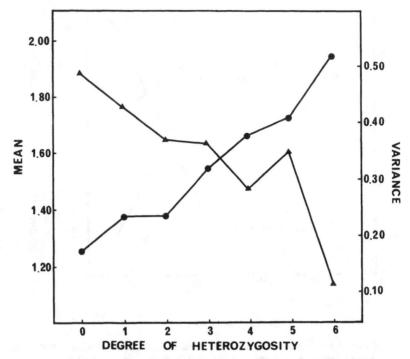

Figure 5.6. Mean and variance of the ln transformed weights of the oyster, *Crassostrea virginica*, of seven classes each containing individuals with the same number of heterozygous loci. Only individuals scored for seven loci are included. Circles, means; triangles, variance. (Source: Zouros, Singh, and Miles, 1980.)

average heterozygosity of each class measured at 27 electrophoretic loci. It was found that average heterozygosity increases with increasing age class, whereas the genetic variance decreases. In addition, a positive correlation was found between individual heterozygosity and fecundity, longevity, and speed of development. Both age and reproductive potential were positively correlated with individual heterozygosity, and plants flowering after 2 years were more heterozygous than plants which did not flower. Schaal and Levin concluded that individual heterozygosity is being selected for and could serve as a powerful buffer against demographic events that might lead to a decline of polymorphism as may occur under periods of environmental harshness, since in these circumstances extreme-environment heterosis would be manifested. Because *Liatris* is a structured population, this is a particularly favorable situation for the observation of heterozygote advantage (Frankel and Soulé, 1981). Most of the organisms discussed in this book are far more vagile, being animals. In a structured population, rather large arrays of linked alleles can be maintained in nonrandom or nonequilibrium populations for several generations (Wills, 1978). This means that structured populations will produce a much greater range of individual heterozygosities, compared with the situation expected in more vagile species. In such populations, there are good chances of finding evidence of heterozygote superiority when assaying a random sample of loci because even apparently neutral alleles will often be linked to genes conferring an advantage when heterozygous.

An additional plant example is a hexaploid *Veronica peregrina*, where electrophoretic banding patterns were found to be more variable at the edges of vernal pools than at their center (Keeler, 1978). Water availability and temperatures fluctuate more at the edges than in the center. In addition, competition is certainly less predictable at the periphery since there is a variety of competitors, whereas at the center of the population, competition is mainly intraspecific. This result follows on from Linhart (1974), who found parallel differences for morphological traits, so that the variation patterns at the morphological and molecular levels are similar. The distance separating the center from the periphery is only 2 to 5 m, and gene flow undoubtedly occurs. It is over this distance that the selective regimes in the two environments must differ quite substantially.

Turning to interpopulation studies, Garten (1976) investigated the relationships between aggressive behavior and genic heterozygosity in the Oldfield mouse, *Peromyscus polionotus*, and found a positive correlation between mean heterozygosity and measures of aggressiveness, social dominance abilities, and ability to compete successfully for limited food. Body weight was positively correlated with measures of social dominance, food control, and aggressive grooming (and, hence, heterozygosity levels). Garten suggests that in peak and declining populations of small mammals competition for space and resources will put nonaggressive phenotypes at a disadvantage, so that in a population crash mostly aggressive heterozygous individuals survive. In addition, he argues that fluctuating uncompetitive and competitive environments may give rise to density-dependent selection for heterozygosity.

Soulé (1979) attacked the problem in a different but related way (Figure 5.7). He took asymmetries between right- and left-hand sides as indicators of developmental stability or homeostasis. Since environmental stress is known to affect the asymmetry of bilateral structures, if heterozygosity is associated with overall fitness, then the most heterozygous individuals should show the lowest asymmetry levels. For 15 populations of the side-blotched lizard, *Uta stansburiana*, the correlation between a measure of morphological asymmetry and percent heterozygosity at 18 electrophoretic loci came to -0.55 ($df = 13$,

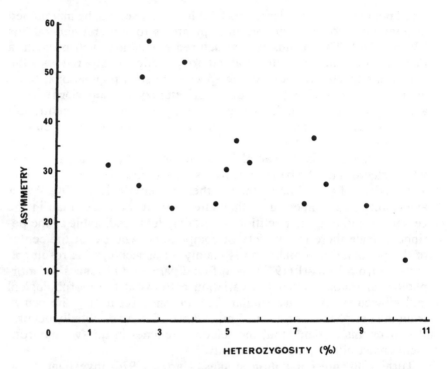

Figure 5.7. The relationship between total asymmetry and percentage heterozygosity in 15 populations of the side-blotched lizard, *Uta stansburiana*. (Source: Simplified from Soulé, 1979.)

$P < 0.05$). He comments that "if further studies verify this result, we will be led to conclude that individuals from relatively heterozygous populations are fitter by this criterion than are individuals from relatively homozygous populations." He makes the point that comparisons of this nature are only possible within species – a point to be developed in the next chapter.

There are two major conclusions:

1. In outbred animal populations there is an association between heterozygosity levels and fitness measured in a variety of ways. Although few plant examples have been cited, the conclusion certainly applies to some plants (Frankel and Soulé, 1981). For example, the presence of segregating polymorphisms in some populations of predominantly inbreeding species has been interpreted as evidence for heterozygote superiority; in experimental populations of barley, *Hordeum vulgare*, the fitness of heterozygotes must be twice as high as that of the corresponding homozygotes to maintain these polymorphisms in the face of intense inbreeding (Clegg, Allard, and Kahler, 1972). How-

ever, in inbreeding plants, there is a range from populations
that are completely homozygous to populations where heterozygotes are in excess, whereas in between are populations
where there is some heterozygosity without evidence for heterozygote superiority (Brown, 1978; Frankel and Soulé, 1981).

2. In populations from marginal habitats, which are likely to be
    stressful, heterozygosity is likely to be higher than for popu-
    lations from more benign habitats. As already commented,
    there is some evidence for this for electrophoretic loci in nat-
    ural populations of *D. robusta* and *D. melanogaster*. Interpre-
    tations become difficult since in marginal habitats small foun-
    der population sizes may counter this tendency. However, if
    heterozygote advantage is common in marginal habitats, the
    reduction of variability predicted from small population sizes
    would be less than predicted on neutral gene calculations of
    the type carried out by Nei, Maruyama, Chakraborty (1975);
    this conclusion could apply to isofemale strains in the labo-
    ratory (Section 4.4). It is important, however, to reiterate that
    extreme populations may show low variability for ecological
    phenotypes because of continuing directional selection due to
    the extreme conditions (Section 4.3) without any associated
    fall in variability for electrophoretic loci.

5.5 Summary

There is a substantial literature on variation in central and marginal
populations based upon gene and chromosome polymorphisms. Most
data show a reduction of chromosome polymorphisms and a reduction
of lethals and semilethals toward the margins, but no equivalent reduction in electrophoretic polymorphisms. Since these are genotypic
assessments not directly relatable to the field situation, the somewhat
unsatisfactory nature of some of these data is understandable.

By contrast, it is important to consider the organism as the unit of
selection. This is illustrated by ethanol tolerance in *Drosophila*, which
can be regarded as an ecological phenotype related to resources in
nature. Contrary to what would be expected, the association of this
trait with alleles at the *Adh* locus and with ADH enzyme activity appears not to be high in *D. melanogaster*. This stresses that ecological
phenotypes should often form the basic data in the analysis of populations, which is an approach differing from that of commencing with
electrophoretic polymorphism data.

The analysis of ecological phenotypes is made complex, since the
amount of underlying genetic variability expressed depends upon the
environment. In particular, it appears that additive genetic variability

is maximized under conditions of extremely hard selection as would be imposed from extremes of temperature and desiccation. From this, it follows that colonizing populations may be relatively simple to analyze genetically, because of their tendency to experience severe environmental stresses, even if sporadically.

Although electrophoretic polymorphism levels are often difficult to interpret, there are an increasing number of studies indicating associations between heterozygosity levels and fitness measured in a number of ways, especially when conditions are stressful. High heterozygosity levels are, therefore, likely to be preserved during colonization events, unless founder populations are so small that bottleneck effects occur.

6

Colonizing phenotypes and genotypes

> If therefore an organism be really in any high degree adapted to
> the place it fills in its environment, this adaptation will be con-
> stantly menaced by any unexpected agencies liable to cause
> changes to either party in the adaptation. [Fisher, 1930:41]

6.1 Genetic variability levels among widespread species

The comparative study of congeners is a potentially important approach in understanding adaptations of colonizing species and could
reveal special genetic features of colonists, although many potential
difficulties in this aim will be apparent from the previous two chapters.
At the phenotypic level, there is certainly a parallel between the ecological phenotypes of marginal populations and the characteristics of
colonizing species, by comparison with congeneric noncolonist species
(Chapter 3). Here, the question is the degree to which this phenotypic
conclusion can be applied at the genotypic level.

Commencing with the eight cosmopolitan *Drosophila* species that
occur in the six biogeographical zones, Carson (1965) has classified
them according to degrees of chromosomal morphism:

1. Polymorphism essentially ubiquitous; a few rare local aber-
   rations – *ananassae, busckii, hydei, immigrans,* and *melanogaster.*
2. Polymorphism frequent and extensive in some populations but
   reduced in others – *funebris.*
3. Monomorphism throughout – *repleta, simulans.*

Considering the major subgenera of the genus *Drosophila* (Table 1.2),
subgenus *Drosophila* (*hydei, immigrans, funebris, repleta*) is represented in all three categories, whereas the sibling species *D. melanogaster* and *D. simulans* are in different categories. Hence, there is no
association with taxonomic divergence in contrast with resource-utilization patterns of three sympatric species of the Melbourne area (Section 3.4) and those of a number of species from a market in Leeds,
England (Atkinson and Shorrocks, 1977). Very high levels of inversion
polymorphism, such as found in certain endemic species, for example,

87

D. willistoni, do not appear to be a general feature of cosmopolitan species. Indeed, of the cosmopolitans, only *D. funebris* shows a pattern resembling that of such polymorphic endemic species (Carson, 1965). *D. funebris* is highly polymorphic in large urban areas, such as Moscow, whereas outlying rural areas have very little inversion heterozygosity at any time of the year (Dubinin and Tiniakov, 1947). Collections outside Russia have shown *D. funebris* to be essentially monomorphic. Hence, geographical expansion throughout the world must have involved marginal populations with low polymorphism relative to the center of the endemic range in Russia. Although *D. melanogaster* is described as polymorphic, aberrant chromosomes occur at a relatively low frequency; colonizations may well occur with chromosomal arrangements largely fixed in homozygous condition – a generalization that appears to apply to varying degrees to all cosmopolitan species.

One other very widespread species on a worldwide basis is the resource specialist, *D. buzzatii*, that exploits the cactus genus *Opuntia*. It is American in origin and has spread with *Opuntia*. Populations from Argentina carry several inversions. In this case, levels of inversion polymorphism in colonized areas are comparable with those in the ancestral area. From a lack of correlation between biogeographical conditions and polymorphism levels, Fontdevila et al. (1981) suggest that macroenvironmental factors must be of low importance. Additional observations on these and other species will be of considerable interest, given that widespread species are increasingly being found considerable distances from their centers of origin, for example, *D. subobscura* in South America (Brncic and Budnik, 1980) and *D. pseudoobscura* in New Zealand (Parsons, 1981a). Founder events may, of course, be of considerable importance in determining the genome of these "remote" colonizations, and studies of ecological phenotypes are needed in all cases.

Therefore, the expectation of low levels of inversion polymorphism in cosmopolitan (and other widespread) species appears generally true. When, however, gene polymorphism is considered, the situation is more difficult to interpret. For example, the level of electrophoretic polymorphism is higher in *D. pseudoobscura* than in *D. persimilis*, which is higher than in *D. miranda* (Prakash, 1977b), paralleling the range of geographical distribution of these three *obscura* group species. However, there are indications that there is less electrophoretic polymorphism in *D. busckii* and *D. buzzatii* compared with other widespread species (Prakash, 1973b; Barker and Mulley, 1976). These authors consider the possibility of founder effects and drift but argue that the low levels of heterozygosity are adaptive, reflecting narrower niche widths relative to *D. melanogaster* and *D. simulans*. Since ecological

Table 6.1. *Proportions of loci polymorphic, average number of alleles per locus, and proportion of loci heterozygous for 14 electrophoretic loci in four anuran amphibians of Israel*

Species	Proportion of loci polymorphic	Average no. of alleles per locus	Proportion of loci heterozygous
Pelobates syriacus	0.11	1.11	0.052
Rana ridibunda	0.44	1.55	0.088
Hyla arborea	0.50	1.86	0.088
Bufo viridis	0.56	1.84	0.169

Source: Data from Nevo (1976).

interpretations of this nature are difficult to substantiate except for the resource-specific *D. buzzatii*, explanations advanced can only be regarded as tentative at this stage (Lewontin, 1974). Equivalent difficulties apply to plants (Jain, 1979).

On the other hand, four anuran amphibian species from Israel occupy increasingly variable and unpredictable environments in the following order: (1) the spade-foot toad, *Pelobates syriacus*, a subterranean narrow habitat specialist, (2) the edible frog, *Rana ridibunda*, a semiaquatic and aquatic habitat intermediate, (3) the tree frog, *Hyla arborea*, an arboreal habitat intermediate, and (4) the green toad, *Bufo viridis*, a terrestrial broad habitat generalist. Table 6.1 shows that heterozygosity is positively correlated with environmental heterogeneity and habitat unpredictability (Nevo, 1976). Even though this result suggests that levels of electrophoretic polymorphism are adaptively important, such interspecific comparisons are often difficult to make because of problems in comparing niche breadths. Indeed, after reviewing electrophoretic studies of 243 species of plants and animals, Nevo (1978) concluded that the niche variation hypothesis was supported by overall trends but that further well-designed tests are needed. Probably, such comparisons are only meaningful among predominantly sympatric populations that utilize similar resources. Wider comparisons would appear to be extremely difficult to interpret, and often conclusions appear contradictory at best. In any case, levels of observed heterozygosity are usually quite high and of the same order of magnitude in most diploid organisms whose natural populations have been studied (Johnson, 1979a,b).

6.2 Frequency- and density-independent selection?

Central populations, being in the area ecologically most favorable for a species, tend to be large. According to Clarke (1979), the major factors affecting population size are predators, parasites, and competitors, all involving frequency dependency. He contends that (1) predators that concentrate disproportionately upon common varieties of prey overlook rare ones, which, therefore, increase, (2) frequency-dependent selection by parasites could be important in maintaining biochemical diversity in their hosts, and (3) frequency-dependent interactions occur when members of the same species compete for a limited resource (including potential mates). Therefore, he considers that frequency dependency is a powerful force in maintaining genetic diversity.

In ecologically marginal habitats, however, direct climatic selection may reduce the importance of frequency dependency. The same is likely in highly variable climates, where, in phases (which may be quite short) of extreme environmental stress or resource limitation, selection is severe, leading to quite dramatic changes in the distribution and abundance of species (Wiens, 1977). For example, during 1975–7 there was a major drought in California. Ehrlich et al. (1980) used this opportunity to look at the dynamics of two species of the checkerspot butterfly, *Euphydryas*. Several populations became extinct, some were dramatically reduced, others remained stable, and at least one increased. The different responses can be related to the fine details of the relationship between insects and their host plants, which vary among populations. The extinctions show that such events may not be uncommon, but because of their unpredictability, detection is difficult. Ehrlich et al. in fact suggest that such severe climatic stresses may occur once only every 50 to 100 generations in California, so that long-term studies are essential to pick up stress phases.

In flowering plants, direct evidence of such catastrophic selection consists of the observation that ecologically marginal populations of several species of the western American annual of the genus *Clarkia* (Onagraceae) have suddenly become extinct under conditions of exceptional drought (Lewis, 1962). The geographical limits of distribution of a given species within the region occupied by the genus are determined primarily by its ecological range of tolerance. Since the climatic factors limiting the distribution of a species occur intermittently, marginal populations may become periodically extinct. In *Clarkia*, it is argued by Lewis (1962) that catastrophic selection due to environmental extremes may sometimes leave one or more exceptional individuals that may found populations characterized by deviant genomes, themselves perhaps attaining the status of species during recolonization phases following extinctions.

An extraordinary view of environmental variation discussed by Wiens (1977) is that provided by Wigley and Atkinson (1977), who used precipitation records gathered at Kew, England, from 1698 to the present to calculate deficits in soil moisture during growing seasons (Figure

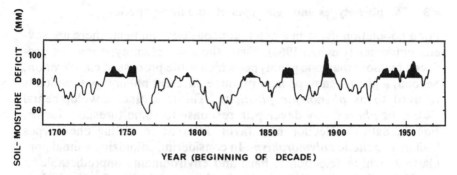

Figure 6.1. The soil-moisture deficit at Kew, England, has been averaged over the growing seasons of a 279-year period. The plotted values are 10-year running means, and, thus, the value for any given year is the average deficit for that year and the 9 following years. In this analysis, then, severe soil-moisture stress is considered in terms of averages over a decade rather than single extreme years. An arbitrary level of 84 mm is shown to accentuate periods of higher deficits. (Source: Wigley and Atkinson, 1977.)

6.1). This measure provides a fairly direct assessment of the degree of drought stress on the vegetation and association fauna. Hence, the measure should be closely tied to productivity and resource levels. Figure 6.1 shows that soil moisture deficits were very variable over the 279-year period. It is particularly noteworthy that there is no apparent regularity in the occurrence of severe years.

As Wiens comments, this analysis is a dramatic demonstration of the extreme variability of climate, and hence the difficulty of predicting severe droughts. These would be times when hard selection (Section 3.1) is maximized so that major effects upon the distribution and abundance of organisms would be very likely, as also emphasized by Nix (1981) from plant biogeographical considerations. Therefore, unless populations are in some sort of equilibrium with their resources, frequency- and density-independent selection is unlikely. Hence, in considering colonizing species, population models invoking frequency- and density-dependent selection can be largely ignored, which includes competition with related species. Price (1980) argues for the insignificance of competition when reviewing the evolutionary biology of parasites; these are organisms with a pattern of repeated colonizing events as will be elaborated in Chapter 11. A future major field of study could well be the comparative adaptations of the colonizing species discussed here and their parasites. Finally, in a recent discussion of population growth, Prout (1980) presents theoretical and experimental arguments for the widespread occurrence of density-independent selection in natural populations.

6.3   The phenotypes and genotypes of colonizing species

For a population living in a heterogeneous environment, there are several outcomes (Levins, 1968). First, the same phenotype may be produced by more than one genotype, which is the process of *canalization*. Second, a given genotype may produce different phenotypes, which is referred to as *phenotypic plasticity*. Third, distinct developmental classes or *phases* may develop in response to some threshold factor. Fourth, natural selection may favor different coexisting phenotypes leading to *genetic polymorphism*. In considering colonizing animal populations, where selection is hard and environments unpredictable, it would be likely that phenotypic plasticity would be of rather minor importance as compared with more benign habitats, where adaptation to regularly occurring minor fluctuations might occur by means of phenotypic plasticity. Thus, for animal colonists at least, population models concentrating largely upon genotypic responses appear appropriate, which provides a level of simplicity paralleled by the assumption that models invoking frequency- and/or density-dependent selection may be relatively unimportant (see, however, Section 11.2 for evidence for phenotypic plasticity in plants).

Remington (1968) considered that the genetic structure of the source population may have profound effects on the chances of an initial colonization being successful, the colony then developing, and in some instances the chances of immigration, but that both practical and theoretical work has tended to ignore genetic aspects. Section 5.1 shows that there is now a large literature on genetic variability per se without much consideration of the ecological phenotype. More recently, however, ecological aspects have begun to assume more importance (Section 5.2). The demonstrated importance of climatic selection in determining the ecological phenotypes of marginal populations provides a starting point for genetic analyses. Additionally, since colonizers tend to be generalists for resource utilization, the genetics of traits, such as ethanol tolerance (in *Drosophila*), is important. An ultimate aim is the development of definitive phenotypic and genotypic predictions concerning the likelihood of species introduced into new habitats for biological control becoming established, or the likelihood of species colonizing new habitats when barriers, such as the Suez Canal (see Safriel and Ritte, 1980), are opened. The ecology of the new habitats is critical in such predictions, especially as colonization events that are demonstrably ecological as in *Dacus tryoni* (Section 2.1) have been rarely witnessed. In this context, Graham, Rubinoff, and Hecht (1971) studied the temperature physiology of the sea snake, *Pelamis platurus*, as an index of its colonization potential into the Atlantic if a Central American sea-level canal is constructed. They concluded that its temperature physiology would be permissive of spread as far north as the English Channel and Cape Cod in the summer months. In the subtropical pierid butterfly, *Nathalis iole*, geographically distinct populations have evolved an ability to anticipate seasonal conditions through an adjustment in melanin synthesis mediated by photoperiod (Douglas and Grula, 1978). In this way, sexual activity is maintained in temporally and geographically cool environments so facilitating a northward ecological range expansion by the midwestern "aggregate." Several other butterfly species also exhibit seasonal polyphenisms (Shapiro, 1976; Hoffman, 1978). More generally, ecological disasters caused by unexpected explosions of introduced species (Elton, 1958) could well have been avoided if an assessment of colonizing potential had been formed in advance.

The approach then is to first investigate how ecological phenotypes fit into the climatic and resource characteristics of their new habitats. A number of studies cited show that certain species do not have ecological phenotypes for major range expansions, whereas other closely related species do. If a degree of correspondence is found between the new and old habitats at the interspecific level, then intraspecific adaptations of ecological phenotypes may occur by natural selection, or in the case of an organism of economic significance, breeding procedures may also be employed. At the applied level, the prerequisite for a successful introduction is, therefore, an appropriate ecological phenotype at the interspecific level that is then potentially modifiable intraspecifically. In any case, a colonist may need to respond quickly to environmental conditions on the edge of, and possibly outside the range of, conditions in its original environment during a range expansion. Knowledge of the range of variation of a phenotype and its genetic architecture at the population level is a prerequisite for predictive purposes.

The data needed to study the evolutionary consequences of invasions, therefore, include:

1. The history of the introduction including population sizes. As shown in Section 4.4, population sizes need to be very low for there to be major consequences for genetic variability levels. Since studies normally begin *after* the colonization event, this information is, however, normally unavailable. Artificial habitat changes, such as the opening of canals, should afford some opportunities for such studies in the future, as should detailed studies of events following the introduction of organisms for the purposes of biological control.

2. The study of the population size expansion and the range expansion following the introduction.

3. The study of ecological phenotypes, and more broadly, life

histories, in the old and new habitats. This includes genetic studies. In some cases, it will also be important to consider behavioral phenotypes, as will be discussed in subsequent chapters.

4. While variability data from electrophoretic loci have turned out to be relatively uninformative as discussed in Section 6.1, such data will provide reference points especially if there are changes in ecological phenotypes.

Data on electrophoretic loci will in any case permit an assessment of the genetic distance between introduced and source populations. Any shifts in gene frequencies after an introduction should be investigated fully, incorporating ecological phenotypes and a demographic–genetic analysis of populations in nature.

As an example where some of the aforementioned data have been collected, 101 marine toads, *Bufo marinus*, were introduced into northern Queensland, Australia, in 1935 and subsequently released in all major sugar cane-growing districts in coastal Queensland during 1935–7. These introductions provided the starting points for exponential geographical expansions (Figures 6.2 and 6.3), which continued for 40 years at 8.1% per year (Sabath, Boughton, and Easteal, 1981). Sabath (1981) has shown that there is a north–south difference in allele frequency at the sorbitol dehydrogenase (*Sdh*) locus, which must have been established in less than 25 generations. The cline is steep (Figure 6.4) and consistent with maintenance by natural selection. However, identifying the type, direction, and strength of natural selection on the *Sdh* locus requires a detailed analysis of the populations in relation to their environment, as well as an extensive analysis of population characteristics. The example of *B. marinus* is cited since it is the type of case study that would be important to seek for future studies, simply because the actual introduction is documented. Unfortunately, appropriate techniques were unavailable for combined genetic and ecological studies in 1935. However, *B. marinus* is still occupying more territory, affording some opportunities for the study of colonization in progress by following the leading edge of the distribution. Eventually, Sabath, Boughton, and Easteal (1981) consider that spread will be limited by temperature and moisture. The southern distributional limit will probably be determined by cold climate affecting developmental rates in both aquatic and terrestrial stages, whereas spread to the interior will be limited by insufficient breeding sites due to lack of water. However, temperature and moisture should not limit the present population expansion across the monsoonal northern portion of Australia.

Another widespread vertebrate that may repay study is the house sparrow, *Passer domesticus*, which is a grain-eating commensal with man. The species is known to have occurred 10,000 to 15,000 years

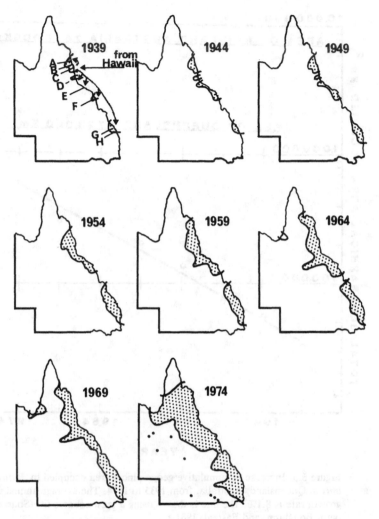

Figure 6.2. Geographical expansion of the marine toad, *Bufo marinus*, in Queensland, Australia, from 1935 to 1974. Sites of original introduction: A, Mossman; B, Gordonvale; C, Babinda; D, Bamaroo, E, Giru and Ayr; F, Mackay; G, Bundaberg; H, Isis. Dots in map for 1974 represent isolated interior records. (Source: Sabath, Boughton, and Easteal, 1981.)

ago in one region of the development of sedentary agriculture based upon wheat and barley. The house sparrow has relatively small population sizes but is capable in each generation of undertaking considerable dispersal movement, so enabling the ready exploitation of available resources. Morphological adaptations that are partly genetic have been found and these may occur quite rapidly. For example, relative

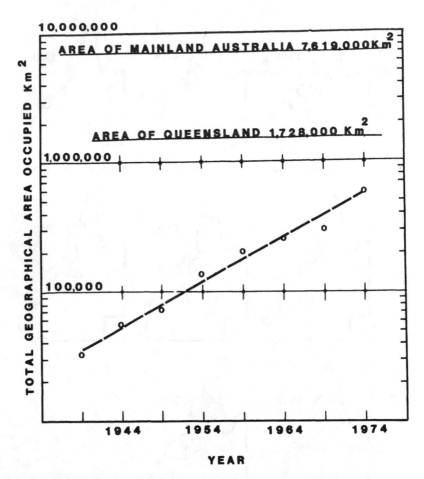

Figure 6.3. Increase in cumulative geographical area occupied by *Bufo marinus* in Queensland, Australia, from 1935 to 1974. The average annual growth rate is 8.1%. The line is drawn using a least-squares fit. (Source: Sabath, Boughton, and Easteal, 1981.)

to English birds, wing length has increased 3 mm in size in not more than 110 generations in North American birds (Johnston and Klitz, 1977). Therefore, upon occupying new habitats, such as North America, a complex series of adaptations may occur in quite short time periods that should repay future studies.

The point has been made several times that one increasingly accepted generalization is that the genetic architecture of morphological and environmental stress traits consists mainly of a rather low number of structural genes with relatively large, mainly additive effects. On the other hand, in some cases regulatory genes are being increasingly hy-

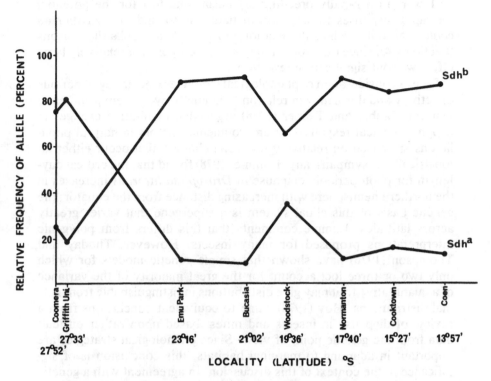

Figure 6.4. Relative frequency of two alleles at the *Sdh* locus in relation to locality latitude in Australian *Bufo marinus* populations. (Source: Sabath, 1981.)

pothesized as being of importance in population differentiation (Carson, 1977; Dickinson, 1980). Since regulatory genes may be most important in precise adjustments to a predictable environment, their significance may be lower under the extreme environmental variability to which marginal populations are subjected compared with central populations. It was also argued earlier that phenotypic plasticity is likely to be unimportant in such circumstances. Accordingly, the genetic architecture of the ecological phenotypes of marginal populations may mainly comprise additive genes. Indeed, additive genetic control dominates under an array of extreme environmental stresses compared with more benign environments (Section 5.3). Provided that populations are polymorphic for such genes, there is potential for quite rapid evolution of new races during a colonization episode. A genetic architecture of many rather than few polygenes or where interactions among genes are widespread (as in inversions) would have more resilience, whereby rates of response to a given selection intensity would be lower. In a rapidly breeding organism, one test for the potential for rapid responses to selection in nature is to find out whether an ecological trait, such as desiccation in *Drosophila*, tracks the seasons (Section 4.4), since this would imply rather few genes of relatively large effect without significant interactions.

From a consideration of phenological strategies (the timing of periods of activity and dormancy in relation to annually cycling environmental factors), Tauber and Tauber (1978) argue that qualitative changes in ecophysiological responses to environmental factors in natural populations are based on relatively few gene changes that occur either allopatrically or sympatrically. Lumme (1978) found that the critical daylength for photoperiodic diapause in *Drosophila littoralis* increases in the northern hemisphere with increasing distance from the equator; the genetic basis of this clinal system is a supergene that varies greatly across latitudes. Lumme comments that this differs from polygenic interpretations proposed for many insects. However, Thoday and Thompson (1976) have shown that simple genetic models for which only two or three loci account for the great majority of the variance of a quantitative trait may give distributions indistinguishable from normal distributions. Hoy (1978) came to equivalent conclusions from a review of diapause in insects and mites, based upon rather difficult data from the genetic point of view. Since phenological strategies are important in adapting to marginal habitats, this conclusion is of significance in the context of this discussion. In agreement with a genetic architecture based upon relatively few genes are substantial responses to directional selection in the laboratory for critical day length in *D. littoralis* and other insects (Lumme, 1978). Conclusions are, therefore, similar from both field and laboratory data.

6.4   The Australian Aborigine – a case study

Humans were established in southern Australia at least 35,000 years B.P. (Bowler, 1976), and possibly up to 40,000 years B.P. However, Thorne (1977) considers that the variation in human fossil material is consistent with occupation at much earlier times. Undoubtedly immigration would have been assisted by lowered sea levels, which were at their lowest about 50,000 and 20,000 B.P. The settlement of the continent was undoubtedly far from uniform. For example, Bowdler (1977) suggests that "the continent was colonized by sea, by people with a coastal economy which underwent little modification for many millenia. Such modification in the initial stages merely involved a shift from marine to freshwater aquatic resource exploitation, and areas away from the coasts and major river/lacustrian environments were unpopulated until rather late in the day." An initial peripheral occupation by the colonists is, therefore, proposed.

Much has been written on the origin and racial composition of the Aborigines. Two main theories have been put forward in the literature (Kirk and Thorne, 1976):

1. The Aborigines are a homogeneous population with no significant variation.
2. The Aborigines are a product of hybridization between two or more races, probably three. These races are supposed to have migrated separately to Australia, followed by hybridization within Australia.

The proponents of the homogeneity theory essentially propose that any differentiation among the Aborigines that cannot be attributed to environmental factors, mutation, drift, or selection is within the range of variation of a single racial group (Parsons and White, 1976). By comparison, those adhering to the hybrid theory believe that the variation can only be explained by invoking more than one race. Because the traditionally oriented Aborigine is now largely extinct, it is difficult or impossible to confirm either hypothesis. A semantic problem is the issue of the amount of divergence between two human populations that constitute a race. The term *race* is unfortunately arbitrary, since there are no biological isolating mechanisms between human populations that characterize different species. It could be said that populations differing by a certain arbitrary amount measured in terms of standardized genetic differences represent distinct races. However, even the calculation of genetic distances themselves is arbitrary, depending upon the loci and/or the anthropometric measures used (Ehrman and Parsons, 1981a). Therefore, heterogeneity between different Aboriginal groups does not argue for either of the hypotheses without historical evidence of one or separate waves of migration of different races from outside Australia.

In considering the initial colonization and occupation of Australia by the Aborigine, it is important to consider population size variations. This is because of the possibility of founder effects during colonization phases and subsequent population crashes. Indeed, catastrophe was a fact of Aboriginal life (Jones, 1977), and its social and biological effects on their history have not yet been fully appreciated. For example, in the Kaiadilt of Bentinck Island in the Gulf of Carpentaria (Figure 6.5), population crashes may have been frequent leading to founder effects, as would be reasonable for a population existing purely by hunting and collecting. Tindale (1962) records that a combination of a tidal wave salting their waterholes, drought, warfare, and sea voyage accidents reduced their population by about 50% in 2 years, whereby the survival

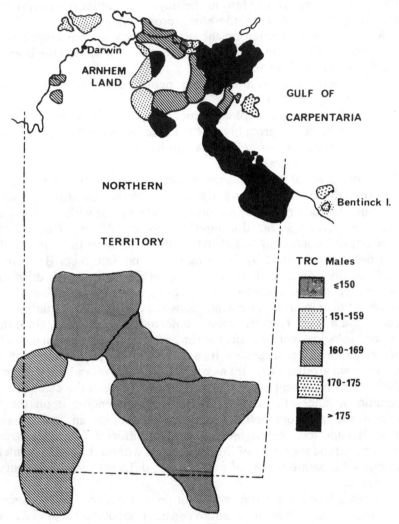

Figure 6.5. Finger total ridge count (TRC) for various tribes in the Northern Territory of Australia (see Figure 1.1). [Source: Updated from White (1979a) incorporating additional data provided by him.]

of the remaining 70 individuals was doubtful until European intervention. Meehan (1977) has documented how a single freshwater flood almost wiped out a major molluscan food resource over a 12-km front of open coast of Arnhem Land (Figure 6.5). Jones (1977) documents other examples where local groups can become rapidly extinct due to the vagaries of the demographic structure. Such population crashes could well have lead to founder effects making it difficult to relate with any certainty the Aborigines of today with other peoples outside Australia. Using blood groups, the gene frequencies of the Kaiadilt appear to have been largely produced by genetic drift (Simmons, Tindale, and Birdsell, 1962). Interpretations remain difficult without knowing the precise history of colonization including numbers at the time of settlement, which are, of course, unobtainable.

A reconstruction of an Indonesian male, *Homo erectus*, revealed a suite of features that suggests regional morphological continuity in Australasia during the middle to late Pleiostocene (Thorne and Wolpoff, 1981). A morphological cline was established through a comparison with males from Kow Swamp, the late Pleistocene Australian site with the largest number of hominid specimens. Interestingly, associated with regional continuity is lesser phenotypic variability in the more peripheral populations compared with East African samples of *H. erectus*. Assuming that the peripheral populations are also ecologically marginal, this corresponds to the low variability of ecological phenotypes in stressful habitats discussed in Section 4.3 for *Drosophila melanogaster*. Such a process suggests the possibility of responses to habitat differences, although drift cannot be excluded if founder population sizes were small during the migration of *H. erectus* out of Africa.

In order to come to an assessment of events after arrival in Australia, one approach is to look at the structure of populations of traditional Australian Aborigines in so far as it can still be assessed. From this it can be asked whether anything can be inferred from the present populations, and their variation that may help to explain past events.

Before European settlement, the Aborigines were subdivided into tribes that were relatively discrete, the isolation being maintained by low migration rates between them. As Jones (1977) writes:

> Australian and Tasmanian Aboriginal tribes are probably best seen as agglomerations of land-owning local groups, who shared the same language and culture, tended to intermarry with each other, met together periodically for ceremonial and other social purposes and foraged on each other's lands to a greater extent than they did with bands of neighboring tribes. Although there was some fluidity in their membership, tribes were important linguistic, genetic and social units.

Birdsell (1973) (and elsewhere) has discussed tribal size among the Australian Aborigines in some detail. Many tribes having a population size varying around 500 are dialectical tribes with no political organization. For mainland Aborigines, Birdsell has shown that population density was proportional to bioproduction as measured by rainfall. This means that there is an inverse correlation between tribal size and rainfall assuming a mean size around 500. Thus, in the richer coastal areas, there were more tribes per unit area, and in the desert fewer tribes per unit area. Jones (1977) comments that "units of this order of size and complexity can be seen as having satisfied such powerful sociological needs that their integrity was maintained, transcending variations in the environment."

Regrettably, there are only a few tribal groups remaining that have sufficiently preserved their traditional hunter–gatherer mode of life permitting meaningful genetic analyses relating to this mode of life. There are, however, a number of Australia-wide analyses of variation of both gene frequency and morphological traits (see Kirk and Thorne, 1976). Considering dermatoglyphics, for example, it was not until Robson and Parsons (1967) and later White and Parsons (1973) that any attempt was made to interpret the observed dermatoglyphic heterogeneity within and between tribes. Perhaps the most detailed research program has been carried out in Arnhem Land to the east of Darwin in the Northern Territory of Australia (Figures 1.1 and 6.5), in particular those tribes still occupying traditional lands (White and Parsons, 1976; White, 1977, 1979a). Apart from investigating population affinities among the Aborigines of this region, an attempt was made to elucidate the possible causes of any differences between the various populations. This involved the following: (1) a consideration of the possible role of culture in determining population genetic structure, (2) the correspondence among populations defined socioculturally, linguistically, and geographically in relation to the biological group, and (3) the degree to which the observed marriage patterns are reflected by the pattern and degree of phenotypic variation. In addition, the degree to which the association of population relationships assessed by dermatoglyphics is paralleled by associations based upon gene frequency and other anthropometric traits was considered (White, 1979a).

The significant findings of the biological program especially the fingerprint survey are summarized in White (1979a). Considering four Arnhem Land tribes, a decrease in genetic relationship (assessed by gene-frequency, dermatoglyphic, and other morphological traits) with increasing geographical distance was found (Table 6.2), suggesting that random genetic drift and/or migration are important mechanisms in producing the observed genetic heterogeneity among the tribes surveyed. A strong association had previously been found between linguistic and genetic features for tribes in the Arnhem Land region (White and Parsons, 1973) and is reinforced in Table 6.2 (White, 1979a). One interpretation is that linguistic drift has occurred within a genetically homogeneous population, leading to the isolation and subsequent genetic differentiation of those populations already separated by language, perhaps as a positive feedback system.

More recent dermatoglyphic studies have concentrated upon the

Table 6.2. *The Spearman rank correlation (r_s) among various "distance" measures computed from four tribes of Arnhem Land, Australia*

	GEOG	LING	GENET	D^2 (males)	MSD (males)
LING	0.66	—			
GENET	0.83	0.60	—		
D^2 (males)	0.77	0.72	0.77	—	
MSD (males)	0.77	0.72	0.77	1.00	—
RD (males)	0.99	0.57	0.81	0.81	0.81

Note:
The number of pairs of comparisons used in each computation was six. GEOG = distance in miles on a map between localities; LING = linguistic distance calculated by subtracting the percentage "common Australian" content of the respective tribes one from another; GENET = distance based on gene frequency traits – blood groups and PTC tasting; D^2 = generalized distance using a modified form of Mahalanobis's D^2 statistic based upon quantitative dermatoglyphic data; RD and MSD = rank distance and modified mean-square distance, respectively. They were devised by White (1979a) as attempts to compute in a simple manner single-distance estimates from mixed categoric and quantitative data. Additional explanations of these distances appear in White and Parsons (1973) and especially in White (1979a).
Source: Data from White (1979a).

broad Yolngu cultural and linguistic complex of northeastern Arnhem Land. These studies have revealed a network of genetically distinguishable and largely intramarrying units. These appear to be related to a number of topographic features including drainage divisions. Indeed, White (1978) has recently emphasized the relationship between genetic diversity and ecological features in Aboriginal Australia. For example, he considers that the inter- and intratribal marriage patterns appear to be strongly influenced by habitat type both at the tribal and the local group level. In turn, these patterns are associated with demographic factors, such as population density and tribal size, both of which are influenced by the physical environment. Quoting from White (1979a), "In Arnhem Land for instance, coastal territories provide a more favorable food economy for hunter-gatherers, with considerable resource diversity in space and stability through time. These coastal tribes tend to be smaller, with higher population densities and shorter marriage distances than those tribes living inland, particularly those in the arid interior." White (1979a) further considers that "it is precisely under the conditions found in Arnhem Land (smaller, more stable population units with marriage systems characterized by polygyny) that we would expect to find the most diverse variation." Figure 6.5 shows this to be true of dermatoglyphic features, and White (1979a) has presented parallel data for cultural features as assessed by language families.

This type of genetic heterogeneity both at the inter- and intratribal level appears, therefore, to be associated (even if weakly) with sociocultural and ecological features. The hypothesis under test is, therefore, that after a sufficiently long period of time, colonists including human hunter–gatherers may become adapted to particular habitats and that this may ultimately be reflected in their gene pools. Much more work needs to be carried out to assess the extent to which environment, that is, habitat, is an important determinant of population structure. However, recent work by S. Sinclair (see White, 1978) using cranial data suggests that a similar relationship existed for Aborigines living along the Murray River (on the border of Victoria and New South Wales, Figure 1.1). It should again be stressed that a lack of knowledge of the history of the colonization events is restrictive in arriving at definitive interpretations.

In Bass Strait between Tasmania and Victoria (Figure 6.6) are a number of continental islands. They have been periodically joined to and separated from their parent continental land mass by glacio-eustatic sea-level changes. The expectation is that the fauna of such islands will be depauperate compared to that of the parent land mass, but unless enough time has elapsed they will be rich in species by comparison with similarly situated oceanic islands, whose species recruitment has been effected entirely across water (Darlington, 1957). The relative richness of continental islands will only be temporary, for eventually the number of species will approach an equilibrium dependent upon geography rather than history. Jones (1977) argues that the same biological forces should apply to hunter–gatherer populations, such as Australian Aborigines. However, several historical conditions must be met. "The men concerned must have watercraft of such limited performance that their colonizing ability across water is negligible; they must be participants in the drowning of a continental shelf where the patterns of final island sizes and interisland distances are sufficiently diverse for the formulation of general statements; and there must be enough time in the new state, for consumation of the effects of biological forces" (Jones, 1977). The postglacial drowning of southern Australia fulfills all of these conditions. Since humans leave archeological records, it is possible to investigate past activities. At the time of ethnographic contact, the Bass Strait Islands (Figure 6.6) were unoccupied; however, there were people on Tasmania itself. Tasmania was joined to the mainland until about 12,000 years BP when Bass Strait became an isolating barrier as it is today. At ethnographic contact a few islands close to the Australian mainland and to Tasmania were used by Aborigines (Figure 6.6), but the more distant ones, in particular

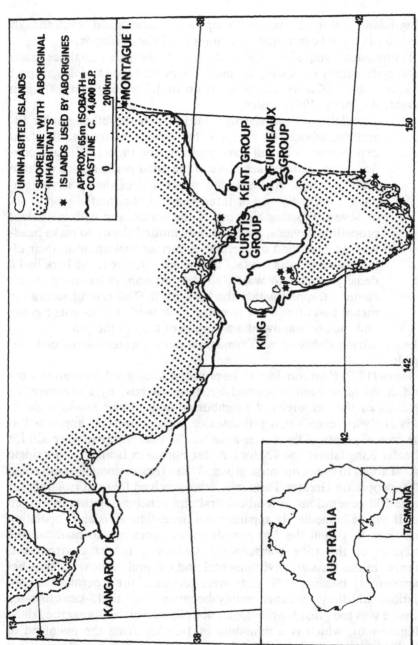

Figure 6.6. Map of southeastern Australia and Tasmania showing the differential Aboriginal use of the various mainland and island coastlines at time of ethnographic contact. (Source: Jones, 1977.)

Legend:
- ◯ UNINHABITED ISLANDS
- ▨ SHORELINE WITH ABORIGINAL INHABITANTS
- ∗ ISLANDS USED BY ABORIGINES
- — APPROX. 65m ISOBATH = COASTLINE C. 14,000 B.P.

0   200km

Labels on map: MONTAGUE I., KENT GROUP, CURTIS GROUP, FURNEAUX GROUP, KING I., KANGAROO I., AUSTRALIA, TASMANIA

King Island and the Furneaux Group, were unoccupied, even though stone tools have been found on a number of these islands.

Tasmanians living along the western and southern coastlines made watercraft usually consisting of bundles of paper bark (*Melaleuca* spp.) or stringy bark (*Eucalyptus obliqua*) up to 4.5 m long and 400 kg in weight. As Jones (1977) writes,

> Calculations from specifications within the ethnographic literature, laboratory tests on the various materials, and field experiments with full-sized replicas [see Figure 6.7] have confirmed eye witness accounts of the good performance of these craft in rough, rock-snagged water and of their carrying capacity, normally two or three people but potentially up to six or seven. Floating high out of the water, and with inefficient propelling devices, they had only limited ability to make headway against wind or current. A further limitation on their effective range was the fact that when saturated, the bark had a density similar to water, so that buoyancy depended on air cavities trapped within the bark itself. The rate of saturation meant that after a few hours, a craft tended to lose its rigidity and thus to wallow like a bundle of kelp in the sea.

The colonizing ability of the Tasmanians across water was indeed negligible.

Jones (1977) then considered the possible Aboriginal population sizes that could have been supported by these islands, by a comparative analysis of the resources of neighboring coastal and inland regions. This analysis reveals that estimates of population sizes supported at the critical points of their separation as the sea level rose were 320 for Greater King Island and 420 for Greater Furneaux Island, itself a single island about to break up into a group. These figures compare with some 4000 people on Greater Tasmania, who survived there in isolation for some 500 generations in total cultural and genetic isolation from then on. It will be immediately apparent that the estimated island population sizes are less than the 500 postulated as necessary to maintain the integrity of the tribe transcending variations in the environment. Hence, in the Bassian environmental and cultural context, islands became empty because their sizes were too small for a permanent population, and they remained empty because a 10- to 15-km cross-sea voyage was too great for recolonization. Similarly, on Kangaroo Island (Figure 6.6), which is a minimum of 14.5 km from the mainland of South Australia, there was no large-scale continuous postglacial occupation for any appreciable time, but only the possibility of sporadic, rare, and relatively short occupations. These situations, therefore, represent rare cases of decolonization that appear consistent with expectations based upon considerations of island biogeography.

Figure 6.7. (*Top*) Tasmanian watercraft, originally drawn in 1802 during N. Baudin's expedition in southeastern Tasmania. (Source: From Lesueur and Petit 1812 Atlas, Plate XIV: accompanying *Voyage de Découvertes aux Terres Australes, 1807–16*, by F. Péron and L. Freycinet, Paris; adapted from Jones, 1977.) (*Bottom*) Reconstruction of Tasmanian watercraft made from *Melaleuca* paperbark by Rhys Jones with the aid of Charles Turner and Ron Wanderwal in 1974. (Source: Jones, 1977.)

6.5   Summary

The difficulties in using gene and chromosomal polymorphism data in attempting to understand colonizing species is confirmed from inter-specific comparisons, even though there is some limited support for an association of electrophoretic variability with niche variation. Consequently, it is necessary to return to the organism as characterized by its phenotype as the unit of selection.

In ecologically marginal habitats, phenomena, such as frequency- and density-dependent selection, and phenotypic plasticity to adapt to varying conditions may be unimportant, because of the unpredictability of periods of environmental stress. By way of a partial summary of previous chapters, the data needed to study the evolutionary consequences of invasions are considered and illustrated with the introduction of the marine toad, *Bufo marinus*, into Australia.

The conclusion of the last chapter that additive genetic variability is maximized under stressful conditions is extended to natural populations. It is considered that the genetic architectures of ecological phenotypes important in colonization comprise mainly a few additive genes and regulatory genes may be relatively unimportant.

This chapter is concluded by a consideration of the colonization of the continent of Australia by Aboriginal people. Although knowledge on heterogeneity among different Aboriginal groups is increasing, interpretations are difficult, since the history of the actual colonizations is unknown. In Arnhem Land, tribes in their traditional state differ genetically. These differences are associated with sociocultural features and with ecological features or habitats. Hence, if isolated for a sufficient period of time, hunters–gatherers may become adapted to particular habitats and this may ultimately be reflected in their gene pools. The average population size of an Aboriginal tribe is about 500. Accordingly, the decolonization of the Bass Strait Islands since the flooding of Bass Strait can be explained, because an analysis of the resources of these islands suggests that supportable populations are less than 500.

7

Behavioral variability in natural populations

> Behavior genetics did not exist as a special field until about 1960. Since that time it has grown considerably. [Caspari, 1977:4]

Following the predominantly ecological considerations so far, the aim of this chapter is to look at behavioral variation with particular emphasis on natural populations. Behavior can be regarded as of great importance in evolutionary change, whether considering aspects concerned with mating or improving adaptedness to a new environment. Behaviors relevant to population continuity need to be isolated and studied under a multiplicity of environments. Assuming that measurement is objective, that is, independent of the observer, there are two emphases that are important in the study of behavioral traits. They are the difficulty of environmental control and the importance of learning (Ehrman and Parsons, 1981a). The first emphasis will be discussed in this chapter mainly with *Drosophila* examples. The second will be considered to some extent in the next chapter, being of greatest importance in organisms higher in the phylogenetic series.

7.1 Phototaxis and light intensities

Hirsch and Boudreau (1958) studied phototaxis (movement with respect to light) in *D. melanogaster* in an apparatus consisting of a Y maze (part of Figure 7.1C), one arm of which was exposed to light during tests. Quite rapid responses to directional selection were found for both positive and negative phototaxis, which is reasonable as a heritability in the broad sense was estimated in excess of 0.5. Similarly, responses to selection have been found in *D. melanogaster* (Hadler, 1964) and in *D. pseudoobscura* (Dobzhansky and Spassky, 1969) for phototaxis in more complex mazes (Figure 7.1C). Upon relaxation of selection in *D. pseudoobscura*, convergence occurred almost as rapidly as divergence under selection, showing that phototaxis in natural populations is a trait subject to genetic homeostasis (Lerner, 1968). Spassky and Dobzhansky (1967) found that geographical strains of the sib-

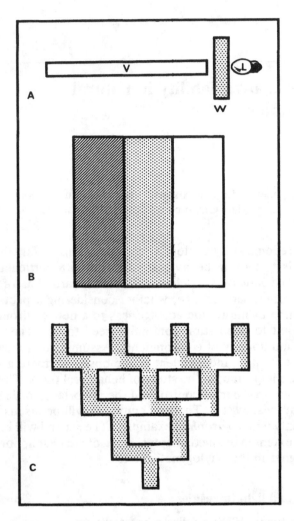

Figure 7.1. Experimental designs for phototaxis analysis in *Drosophila*. (A) Measurement of rate at which flies approach light source. V; Vessel containing flies; W; water-filled heat absorber; L; light source. (B) Measurement of distribution of flies in lighted area. Shaded areas correspond to areas of different intensities of overhead light. (C) Measurement of movement of flies in partly illuminated maze. Unshaded areas represent possible light choices. Lateral movement across several intersections is prevented by a one-way cone inserted in each arm. Illumination comes from above. (Source: Ehrman and Parsons, 1981a.)

ling species *D. pseudoobscura* and *D. persimilis* respond differently in tests of phototaxis. Later, Rockwell and Seiger (1973a) demonstrated the existence of a large amount of variation for photobehavior within populations of each species. Rockwell, Cooke, and Harmsen (1975) corroborated these findings with freshly isolated samples from sympatric natural populations, and found that the level of genotypic variation is higher in the *D. persimilis* population than *D. pseudoobscura* population. A store of variability for phototaxis is apparently available in natural populations, as is the case for almost every quantitative behavioral trait.

Phototaxis also illustrates a rather subtle environmental problem. Under the usual conditions of handling in the laboratory, *D. pseudoobscura* shows positive phototaxis. However, Lewontin (1959) carried out a series of experiments showing that *D. pseudoobscura* is negatively phototactic under conditions of low excitement, but if flies are forced to walk rapidly, or fly, they lose their negative phototaxis and become strongly attracted to light. Rockwell, Cooke, and Harmsen (1975) found that the extent of these environmentally induced changes differs among genotypes from natural populations of this species. Hadler (1964) listed numerous environmental variables affecting phototaxis in addition to those just mentioned – temperature, hour of day when tested, time since anesthesia, rearing conditions, time since feeding, energy and wavelength of light, state of dark adaptation, number of observations or trials per individual, age, and sex. A phototactic response is, therefore, a product of particular stimulus–environment variables as well as of particular genotypes. It is quite clear that any form of accurate genetic analysis must be based on precisely defined environments, as has been repeatedly stressed for ecological traits, but for behavioral traits it is an even more important consideration.

Another complication is the method used in studying phototaxis. At least three different experimental designs (Figure 7.1) have been employed (Hadler, 1964; Rockwell and Seiger, 1973b): (1) measurement of the rate at which flies approach a light source at the far end of a tube; (2) recording, after a specific period, the distribution of flies in a field with a directed or undirected light source; and (3) analysis of movement in mazes. Hadler considers that one of the major difficulties in comparing studies of phototaxis originating from different laboratories is due to interlaboratory differences in experimental technique. Slightly different behaviors may be under study; for one example, the first method confounds phototaxis with photokinesis. Phototaxis is directed movement with respect to a light source; photokinesis is nondirected.

Rockwell and Seiger (1973b) discussed how the various designs differentially accentuate components of the chain of events comprising the totality measured as phototaxis and indicate that the designs differ in their utility for research directed at diverse aspects of the response. For those concerned with the potential adaptive significance and evolution of the behavior, there is no certainty that the response measured in the laboratory is necessarily the same as that occurring in nature. This, of course, is a crucial problem relevant to all laboratory-analyzed behavior.

Although it is clear that all environmental factors influencing either the sign or intensity of the response must be carefully controlled and taken into account in any comparisons, Rockwell and Seiger (1973b) consider that the flexibility of the response relative to environmental changes may be adaptive permitting survival in a heterogeneous environment. For this reason, they propose that in studies concerning the adaptive significance and evolution of phototaxis and other behaviors, responses relative to several values of various environmental variables found in the animal's habitat are as important as responses to a certain set value of each variable.

Therefore, phototaxis illustrates the difficulty of environmental control and its measurement. How then can its relative importance as a component of fitness be assessed?

Oviposition is an obvious fitness trait that is light dependent in a way that appears to depend upon habitat. Allemand and David (1976) analyzed oviposition rhythm in strains of *D. melanogaster* from Africa and Europe (Figure 7.2). Latitudes of origin of the strains ranged from 0° (equator) to above 60° (northern Europe). The experiment consisted of studying oviposition on a 12-hr light (photophase) – 12-hr dark (scotophase) cycle. Figure 7.2 shows that with increasing latitude, daily egg production declined from 80% laid in darkness in flies from equatorial habitats to a little less than 50% from Scandinavian countries. Since the experiments were carried out on strains reared under the same laboratory conditions, the differences must be genetic. Although measurements were usually made upon recently collected wild populations, repeat measurements made after 1 or 2 years of laboratory rearing gave curves very similar to the initial ones, indicating stable genetic differences that persist for many generations as for isofemale strains (Section 4.4). From populations collected over such a geographical distance, it is impossible to assess the direct adaptive significance of the variation in behavior. However, since temperature is strongly influenced by latitude, high correlations of temperature with measures of oviposition behavior were not unexpectedly found.

Returning to phototaxis, studies on natural populations are sparse, although Médioni (1962) found variation for phototaxis in flies from different localities, whereby positive phototaxis increases going north in northern hemisphere samples. Again it is difficult to speculate on

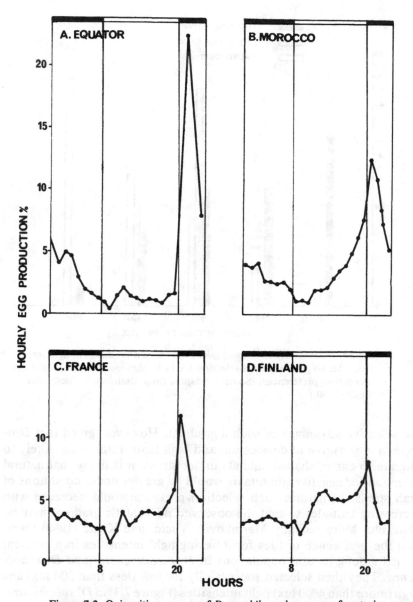

Figure 7.2. Oviposition curves of *Drosophila melanogaster* for strains ranging from the equator to northern Europe. The graphs are for flies from the equator in Africa (A), Morocco (B), France (C), and Finland (D), ranging from a latitude of 0° for A to in excess of 60° for D. The curves are expressed as hourly percentages of the total egg production in a day. The light phase (photophase) was from the 12 hr between 8 a.m. and 8 p.m., and the dark phase (scotophase) between 8 p.m. and 8 a.m. (Source: Adapted from Allemand and David, 1976.)

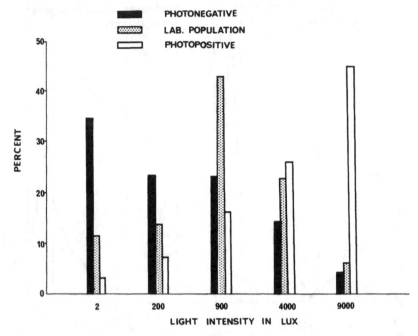

Figure 7.3. Preferences of *Drosophila subobscura* for different light intensities, after 60 generations of selection for high (photopositive) and low (photonegative) preferences. (Source: Adapted from Marinković, Tucić, and Kekić, 1980.)

the selective advantage of such a gradient. However, given that *Drosophila* is sensitive to desiccation and high temperatures and likely to migrate to damp, shaded habitats under stress, it is likely that natural selection for negative phototaxis would be greater under conditions of high stress. In nature, such selection pressure would decrease with increasing latitude, so that the observed phototactic gradient may be plausible. More recently Marinković, Tucić, and Kekić (1980) found that the preference of flies for differing light intensities in a gradient varied among natural populations in *D. melanogaster* and *D. subobscura*. They then selected successfully for low (less than 200 lux) and high (more than 4000 lux) light intensities (Figure 7.3) in *D. subobscura*, clearly indicating that the trait has a genetic basis. Selection for light preference revealed a number of morphological differences in the fine structure of eyes of photopositively and photonegatively selected flies. For example, comparing serially made thick sections of corresponding parts of ommatidia showed that the ommatidia of photonegatively selected individuals are about 30% larger than those of positively selected individuals. Interestingly, in flies from a natural population of *D. me-*

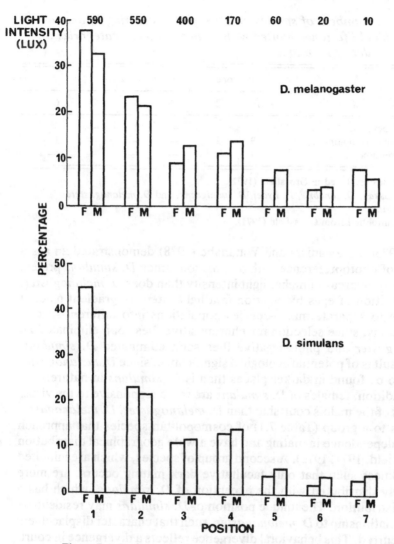

Figure 7.4. Preferences of populations of the two sibling species *Drosophila melanogaster* and *D. simulans* from Melbourne, Victoria, at each of seven light intensities (10–590 lux) in a gradient. F, female; M, male. (Source: Parsons, 1975b.)

lanogaster, individuals that preferred low light intensities had significantly larger eyes than those preferring higher light intensities.

From the aforementioned results, it is not surprising that there may be differences in photopreference between sibling species. For example (Figure 7.4), *D. simulans* shows a relatively higher photopreference in a gradient than does a sympatric population of *D. melanogaster* (Par-

Table 7.1. *Numbers of species of* Drosophila *occurring in one to six geographical life zones plotted with respect to three categories of light dependency in mating*[a]

	Life zones					
Category	1	2	3	4	5	6
Light independent	—	1	—	2	—	5[b]
Facultative dark mating	9	2	—	—	—	1[c]
Dark repression	17	6	—	—	—	—

[a] The species are listed in Grossfield (1972).
[b] *D. ananassae, D. funebris, D. hydei, D. immigrans,* and *D. melanogaster.*
[c] *D. simulans.*
Source: Simplified from Grossfield (1971).

sons, 1975b). Kawanishi and Watanabe (1978) demonstrated an association of photopreference with oviposition, since *D. simulans* prefers to oviposit in areas of higher light intensity than does *D. melanogaster*. The selection of eggs by position in a light intensity gradient made it possible to separate mixed-species populations into different-species populations, since selection for photopositive flies soon eliminated *D. melanogaster* and photonegative flies soon eliminated *D. simulans*. This result is of potential ecological significance, since *D. melanogaster* tends to be found in darker places than is *D. simulans* in nature.

In addition, females of *D. simulans* are more responsive to the visual aspects of the male's courtship than *D. melanogaster*. *D. melanogaster* belongs to a group (Table 7.1) of cosmopolitan species that approach light independence in mating and have a wide geographical distribution (Grossfield, 1971, 1972). A second group of species, which are inhibited by darkness such that only facultative dark mating occurs, are more narrowly distributed with the exception of *D. simulans*, which has a wide distribution. The unique position of *D. simulans* may reside in its close relationship to *D. melanogaster*, such that character displacement has occurred. This behavioral divergence reflects a divergence in courtship organization and its underlying genetic architecture. In view of the latitudinal variations in *D. melanogaster* for phototaxis and oviposition rhythm, comparisons of flies from a number of habitats paying attention to light dependency is desirable, comparing too the sibling species when sympatric or allopatric.

Finally, there is a third group of mainly narrowly distributed species in which mating is repressed by darkness. They include the picture wing Hawaiian drosophilids, where light intensity influences not only the microniche necessary for oviposition but also adult resting positions and courting territories. In these Hawaiian species, therefore, visual

cues are of critical importance, such that particular insects orient themselves in spots of specific and consistent light intensity (Grossfield,
1966). In fact, Grossfield suggested that responses to oviposition stimuli and light levels have contributed to speciation in the Hawaiian *Drosophila*, although subsequently little further work on phototaxis in Hawaiian *Drosophila* has appeared (Carson et al., 1970).

Spieth (1974) comments that in darkness no type of male, mutant or
otherwise, orients normally in courtship. Jacobs (1960) noted that in
darkness abnormal *D. melanogaster* courtships are the rule. For the
effectively light-independent species, therefore, such courtships still
contain sufficient stimuli to lower the female's threshold to the level
at which she will copulate, whereas in the light-dependent species, the
female's threshold remains high. Spieth considers light dependency to
be primitive and that light independence has evolved secondarily in a
number of species, but he finds the adaptive significance of light independence unclear. However, it would be compatible with the life-
style of widespread colonizing *Drosophila* species.

7.2 Physical stresses and behavior

As previously noted, Hawaiian *Drosophila* do not randomly distribute
themselves throughout the forest habitat but select and tend to accumulate in microhabitats meeting their specific requirements. Adults of
most species avoid even moderate wind currents and light intensities,
humidities below 90%, and temperatures above 21°C. So far as the
physical stress components are concerned, humidity and temperature
are likely boundary conditions delineating the extremes within which
responses to other environmental variables occur, for example, light
intensity. In particular, temperature requirements are relatively ex-
acting in Hawaiian *Drosophila* (Carson et al., 1970). In rearing these
species in the laboratory, it is important that the temperature be normally kept under 21°C.

Endemic *Drosophila* in southeastern Australia are found in moist
habitats frequently characterized by tree ferns (Grossfield and Parsons,
1975), where the climate as defined by temperature and rainfall is much
more variable than in the Hawaiian forests. Even so, fern gullies form
refuges, usually isolated from each other, which offer tempera-
ture/humidity conditions compatible with the survival of many *Drosophila* species (Parsons, 1977b, 1981a). Below 12°C, these endemic
flies are apparently inactive and difficult to collect in these habitats.
In parallel, Australian strains of the cosmopolitan species *D. melanogaster* and *D. simulans* do not mate or oviposit below about 12° to
13°C in the winter. Indeed, when flies of either species are placed
between 12° and 20°C, both mating and oviposition increase almost

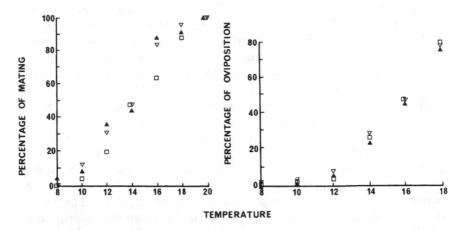

Figure 7.5. Mating percent of three isofemale strains of *Drosophila melano-gaster* after 48 hr at seven experimental temperatures between 8°C and 20°C (left graph), and percent of eggs deposited relative to 20°C at six experimental temperatures between 8°C and 18°C (right graph). (Source: Data from McKenzie, 1975.)

linearly (Figure 7.5) with temperature increase (McKenzie, 1975). In an ectothermic organism, the temperature–energy relation is critical. At lower temperatures, the physiological capabilities of the organism will be limited. It seems reasonable that in winter, energy wastage will be selected against; oviposition during this period represents such a wastage because of the low probability of the development of larvae. Field data for *D. melanogaster* agree, since in the cellar of the Chateau Tahbilk winery in Victoria, larvae and pupae are not found when the winter temperature falls below 13°C although adults can be collected (McKenzie, 1975). This suggests that 12°C or thereabouts may be a lower boundary condition for the mating and activity of flies of at least some of the species of the genus *Drosophila*, although *D. melanogaster* has been shown to survive considerable periods of time at much lower temperatures in a quiescent state (Anxolabehere and Periquet, 1970). Although this conclusion may apply to temperate-zone *Drosophila* species, the narrower tolerances to extreme stresses of tropical species (Figure 3.3) suggests that they should have higher boundary temperatures. Indeed, a boundary of 16° to 17°C occurs in three species collected in tropical Africa, the cosmopolitan *D. ananassae*, and two local species *D. iri* and *D. fraburu* (Cohet, Vouidibio, and David, 1980). In addition, *D. paulistorum* of the tropical Americas is difficult to breed in the laboratory below about 15°C (Parsons, 1978b).

Factors of importance at the interspecific level often turn out to be of importance at the intraspecific level (Chapters 3 and 4). For example, Stalker (1980) studied midwestern and eastern U.S. populations of *D. melanogaster* and found that standard gene arrangement frequencies are higher in the North than in the South. In Missouri, the frequency of standard chromosomes regularly rose during the cold season and fell during the warm season, thus paralleling the north–south frequency differences. Wild males flying in nature in the temperature range of 13° to 15°C showed significantly higher frequencies of standard chromosomes than those in the 16° to 28°C range. Thus, the presence or absence of this common gene arrangement is an important factor in low-temperature adaptation as assessed by flying. Stalker (1980) showed that at least part of the adaptive mechanism involves control of thorax/wing proportions and thus control of wing loading. Extension of studies of this type will be awaited with considerable interest, given the importance of the boundary conditions for resource utilization.

Turning to the upper boundary temperature, it is known that those in the 25° to 30°C range are stressful for many cosmopolitan species although *D. melanogaster* can tolerate higher temperatures than *D. simulans* (Section 3.1). Adaptation to high temperatures and desiccation is a complex of physiological acclimation, genetic variation, and behavior; the relative importance of these three components varies among species (Levins, 1969). Laboratory studies on environmental stresses pay little attention to the mobility of drosophilids, although in a temperature gradient, Prince and Parsons (1977) found that temperature preferences for sympatric Melbourne populations could be ranked as *D. melanogaster* > *D. simulans* > *D. immigrans*, a result not inconsistent with what is known of the ecology of these three species. Behavioral adaptations add a new dimension to the capabilities of an animal, for when stressed it can avoid stress by judicious and alterable microhabitat selection.

Many of the Australian endemic species are collected in the wild by sweeping flies from their actual microhabitats. Because of this, it is possible to characterize the temperature/humidity relationships of the microhabitats. Collection data show quite clearly that flies adjust to microhabitats based on temperature/humidity relationships. Within habitats, temperature gradients occur such that temperatures at the fern frond from which many flies can often be collected are generally in the range 15° to 20°C irrespective of site (Parsons, 1977b). The inferred behavioral responses to temperature and humidity are not surprising, since small insects have an extremely high ratio of surface area to volume, so that the amount of water that can be lost by evaporation is large compared with the amount that can be stored. At higher temperatures, flies become difficult to collect, so that under conditions of extreme stress, genetic differences in resistance to temperature extremes and desiccation may become critical in determining survival (Section 4.3).

As pointed out several times, it is important to investigate intrapopulation variability. McKenzie (1978) approached this problem using seven developmental temperatures over a range of 12° to 30°C for 10 isofemale strains of sympatric populations of *D. melanogaster* and *D. simulans* from Chateau Tahbilk and measured development time, egg-to-adult emergence, adult longevity, mating propensity, and fecundity. He found that isofemale strains of the two species showed similar responses for all traits across temperatures although *D. simulans* was somewhat more affected by extreme temperatures for the latter three traits (see Section 3.1). Even so, the influence of developmental temperature on these life-cycle components was qualitatively similar for both species, as would be expected for an environmental variable of such fundamental significance (Parsons and Stanley, 1981).

Analyses of variance show that temperature, strain, and temperature × strain effects are generally significant (Table 7.2). For percent of matings, Figure 7.6 gives the ordering of three representative isofemale strains of *D. melanogaster* and *D. simulans* showing the same broad pattern at the interspecific level, whereby *D. melanogaster* has a more consistent mating pattern over a broader range of temperatures than *D. simulans*. At the intraspecific level, the ordering of the strains varies among temperatures, explaining the temperature × strains effect. In view of the discussions in Sections 2.1 and 4.3, it is of interest to note that development time gives results analogous to percent of matings, both being frequently assumed to be direct fitness traits.

Isofemale strain variability within a population as in Table 7.2 permits immediate flexibility and changes in population structure under differing and highly heterogeneous environments. This argument is further emphasized by the consistently significant strain (genotype) × environment interactions, which have the effect of increasing flexibility above that from strain differences alone. In conclusion, the differential effects of temperature selection on life-cycle components (including percent of mating) of a spectrum of genotypes may increase the capacity of the population for adaptation in a variable habitat.

The isofemale strain approach is, therefore, of considerable assistance in assaying the role of environments in a general sense in one generation, without the complication of complex crossing procedures. This enables the obtaining of information on the effects of many environments on a series of strains that differ genetically. Since they are derived from natural populations, an idea of the overall responses of natural populations to environmental variables, such as temperature,

Table 7.2. *Summary of analyses of variance of* Drosophila melanogaster *(mel) and* D. simulans *(sim) for various traits following development at a range of experimental temperatures*[a]

Source of variation	Development time		Adult emergence (%)		Longevity Male		Female		Percentage of mating		Fecundity		Sex ratio	
	mel	sim	mel	sim	mel	sim	mel	sim	mel	sim	mel	sim	mel	sim
Main effects														
Temperatures	b	b	b	b	b	b	b	b	b	b	b	b		
Strains	b	b	b	b	b	b	b	b	b	b	b	b	b	b
Replicates														
Interactions														
T × S	b	b	b	b	b	b	b		b	b	b	b		
T × R							b				b			
S × R														

[a] 12°–30°C.   [b] $P < 0.01$.

Source: Data from McKenzie (1978).

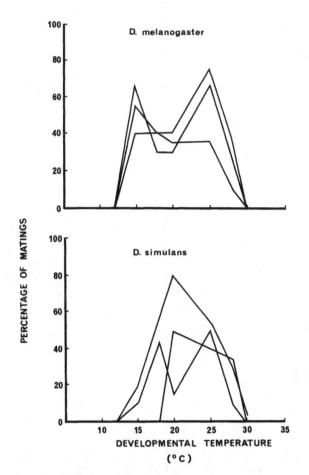

Figure 7.6. The effect of developmental temperature on percent of matings in three representative isofemale strains of the sibling species *Drosophila melanogaster* and *D. simulans*. [Source: Data provided by McKenzie (1978) plotted in Ehrman and Parsons (1981a).]

is, therefore, obtainable. This approach would appear to be particularly valuable for environmentally labile behavioral traits.

7.3   Mating in *Drosophila* (ecology, ethology, and evolutionary genetics)

Mating behavior has been extensively discussed in the literature, often by those with an interest in behavior genetics. More generally, mating in *Drosophila* can be investigated from three points of view: (1) ecology, (2) ethology, and (3) evolutionary genetics.

Ecology

As shown in Section 7.2, the precise environment, in particular, the temperature under which matings occur, is important. Assuming restriction of maximum mating to around the 15° to 20°C range, maximum *Drosophila* activity must occur in temperate regions during relatively short periods of time in the early morning and evening. None of the laboratory mating experiments replicate this field situation, which is desirable to assess the evolutionary significance of variations in mating speed. Predictably, tropical species appear to have a narrower range for activity, mating, and so forth, since the temperature extremes of temperate regions do not usually occur. Even so, temperatures would be in the favorable range for much longer periods of time in tropical compared with temperate-zone rain forests. Intraspecific studies within and among populations are lacking at the present time and would be especially valuable in closely related specialist and generalist species.

Ethology

Bateman (1948) found greater differences among males of *D. melanogaster* for mating success than among females. Presumably these differences have evolved, since male reproductive success is determined by the number of matings achieved, whereas females need only mate once to achieve reproductive success in a given breeding cycle. For two series of choice experiments in *D. pseudoobscura*, $1♀ × 3♂$ and $3♀ × 1♂$, the mean time elapsing to the first mating was 0.53 min in the experiments with three females, and 1.40 min in experiments with three males (Kaul and Parsons, 1966). It appears that in the experiments with three males, competition among males delays mating, whereas in the reverse situation with three females, the one male tends to mate more rapidly having no competition from other males. These results demonstrate the predominant importance of sexual selection among males rather than among females.

It is well known (Spieth, 1974) that the males of some Hawaiian *Drosophila* species patrol and defend a small but definite lek (courting and mating territory). The territories are not randomly distributed but are at specific sites in the vegetation; each species has preferences apparently controlled by ecological factors, such as light, humidity, temperature, and spatial conditions. The territories are close to, but separate from, their breeding sites. Associated with this is the development of sexual dimorphism, and characteristically the species have elaborate picture wings. The existence of such leks will presumably have the effect of enhancing variability among males for reproductive success. More direct evidence for leks (Parsons, 1977c) occurs for two

Figure 7.7. The bracket fungus *Ganoderma applanatum*. This specimen
measures 12.5 cm along the log from which it projects and extends for a
maximum of 10 cm from the log. Flies of the two Australian picture-winged
species, *D. mycetophaga* and *D. polypori*, are found courting on the under-
side of such fungi under suitable climatic conditions. (Source: Parsons,
1977c.)

picture-winged species in the Australian *Drosophila* fauna, namely *D.
mycetophaga* and *D. polypori* (Table 1.2). These flies occur deep inside
rain forests and utilize the undersides of bracket fungi as courting territories (Figure 7.7). The undersides of the fungi are white, light grey,
or light brown, strongly enhancing displays. There may be several flies
that are spread out relatively evenly under an individual fungus. Assuming lek behavior, an excess of males would be expected under the
fungi as is generally found (Parsons, 1978b) in contrast with collections
about the oviposition sites of these flies, which are soft forest fungi,
where, if anything, females are in excess (Table 7.3).

Inside rain forests, such as in Hawaii, periods of high-tempera-
ture/desiccation and low-temperature stresses are minimized (Section
7.2) so that physical conditions may be permissive for the development
of complex behavior patterns, such as lek behavior. A similar release

Table 7.3. *Number of flies of* Drosophila mycetophaga *and* D.
polypori *collected from the undersides of bracket fungi and in the
vicinity of soft forest fungi*

	D. mycetophaga			D. polypori		
	♂	♀	n	♂	♀	n
Bracket fungi	131	27	158	97	53	150
Soft forest fungi	10	12	22	7	13	20
Total	141	39	180	104	66	170
χ_1^2 for independence		13.83[a]			5.35[b]	

[a] $P < 0.001$.   [b] $P < 0.05$.
Source: Updated from Parsons (1978b).

from environmental constraints tends to occur deep inside Australian
rain forests. Indeed, bracket fungi tend to occur in microhabitats with
low light intensities, often close to permanent sources of water. Direct
observation of activity under bracket fungi indicates (1) an upper limit
to resource utilization in the 20° to 22°C range when flies become largely
quiescent under the fungi, and (2) a lower limit in the 12° to 13°C range
when flies become stationary on the fungi and when movement to and
from the surrounding vegetation ceases (Parsons, 1978b). Since these
flies mainly occur in subtropical forests, a lower boundary condition
than for those of tropical forests appears reasonable. Hence, parallel
evolution for lek behavior occurs in Hawaii and Australia in two subgenera (*Drosophila* and *Hirtodrosophila*, respectively, see Table 1.2)
under similar environmental conditions for temperature, humidity, and
light, whereby activities such as feeding and especially breeding are
not restricted to brief diurnal periods.

Given suitable physical conditions, then, leks are occupied. The resources of the Australian species are soft forest fungi, which are ephemeral depending very much upon rainfall patterns in the subtropical regions in which these species are mainly found. In the very richest rain
forests of tropical northern Queensland, where rain occurs on most
days, lek species have not been found. There is, however, a tendency
toward "species packing" for fungus-feeding *Hirtodrosophila* in these
nothern Queensland forests (Parsons, 1981a). Apparently, the lek species occur in habitats permissive in terms of physical conditions for
resource utilization, but because rainfall patterns are unpredictable,
the resources for these species are unpredictable. In this context, Bradbury (1981) working on tropical bats, argues that food dispersion is a
good predictor of the mating system. He argues that classical leks are
established in sites without substantial resources for females and the

emphasis on self-advertisement increases as the resources decrease. The resource unpredictability of the subtropical forests of Australia may, therefore, be conducive to the development of leks.

It can, therefore, be concluded that mating in wild *Drosophila* can only really be understood under conditions where ecological factors are simultaneously studied. There is a clear need to study the cosmopolitan species in parallel ways, since it is upon these species that most evolutionary genetics studies have been carried out. Regrettably, the microhabitats in which the cosmopolitan species mate are in need of much more investigation. However, at the geographical level, in experimental matings between French and Afrotropical populations of *D. melanogaster*, it was found that African males are more sexually active than French males, whereas European females are much more receptive than African females (Cohet and David, 1980). There is in any case often a negative correlation between female receptivity and male mating activity in *Drosophila* (Sturtevant, 1915; Merrell, 1949; Spiess, 1970; McKenzie and Parsons, 1971; Cohet and David, 1980). In the *Drosophila* breeding sites of tropical Africa, up to 10 species of the *melanogaster* species group were observed compared with 2 in France. The greater receptivity of European females may, therefore, have evolved during the spread of *Drosophila* to temperate regions, since in Europe there is a comparatively low risk of interspecific hybridization, and in parallel with this male mating activity declined. Thus, there are variations in sexual selection that may be relatable to habitat.

Evolutionary genetics

There are much laboratory data on the evolutionary genetics of mating in *Drosophila* that have been reviewed elsewhere (for example, Parsons, 1974a; Ehrman and Parsons, 1981a). Typically, inbred strains, mutants, and inversions are used as the study material in laboratory experiments. In summary, there is evidence that

1. Male mating speed is subject to directional selection for rapidity of mating.
2. Within a given species, rapid matings tend to be controlled by the genotype of the male involved, whereas the genotype of the female may assume more importance in slower matings.
3. Mating speed is associated with fertility and number of progeny.
4. Where studied in relation to other components of fitness encompassing the whole life cycle, male mating speed is a most important component of fitness in drosophilids.

Finally, in *D. melanogaster* the offspring of inseminated adult females that mated in a cage where many males were present were found to be fitter as measured by intraspecific competitive success during the part of the life cycle from first-instar larva to adult fly, as compared with offspring collected as virgins and mated with a randomly chosen male (Partridge, 1980). In the former situation, this means that certain males are fitter than others, that is, when there is a selection process.

The aforementioned ethological considerations suggest that male reproductive success must vary much more widely than among females in the wild. In the last 2 or 3 years, several authors have begun to accumulate evidence that this is so in species more extensively studied in the laboratory. For example, (1) Anderson et al. (1979) found evidence for virility selection occurring in natural populations of *D. pseudoobscura* by comparing the frequency of inversions among males and in larvae of females collected at the same time, and (2) Brittnacher (1981) argued that virility may dominate the dynamics of natural polymorphisms of both *D. melanogaster* and *D. pseudoobscura* using males homozygous and heterozygous for the second chromosome of each species.

Assuming this to be so, then *Drosophila* would be analogous to a variety of organisms studied in the field, where male reproductive success varies more than females. Trivers (1972) and others document field examples in a variety of organisms – dragonflies, baboons, frogs, prairie chickens, grouse, elephant seals, dung flies, and certain lizards. The same appears to occur in some primitive human tribes, where polygyny may be an original condition, especially as it is more or less developed in nearly all anthropoid apes. White (1979b; see Ehrman and Parsons, 1981a) demonstrated much greater male fertility differentials than those for females among the Yolngu tribe of the Australian Aborigines of northeastern Arnhem Land (Figure 6.5) in one of the most comprehensive studies of hunter–gatherer populations. These results can be explained by the relative parental expenditure of the sexes on their young. Where one sex, say the female, invests considerably more than the other, then males will compete among themselves to mate with females, as shown in the laboratory in *D. pseudoobscura*.

Conclusions

The evolutionary genetics of the trait mating speed studied in the laboratory is relatively well known and an important component of fitness determined mainly by the genotype of the male. If the results from the lek species can be generalized, then it is likely to be a result of general significance with direct analogies in vertebrates and humans, where individuals can be readily followed. The variability in mating success between French and Afrotropical populations of *D. melanogaster* indicates genetic differences in sexual selection among populations that may be relatable to habitat. Indeed, in outbred populations, evidence is now suggestive of more additive genetic variance for fitness traits than has been conventionally supposed (Section 2.1), so that some genetic analyses of mating speed may be possible.

The lek species, in particular, provide indications of the physical features of the environment needed for mating and, hence, provide the beginning of a link of ecology, ethology, and evolutionary genetics, that is, an ecobehavioral genetics of mating. There is a great need to study the evolutionary genetics of mating under an array of environments especially those simulating the wild. Such an ecobehavioral genetic approach could lead to results applicable to organisms more usually considered in a sociobiological context. One of the major gaps in our understanding of sociobiology is at the level of the underlying genetic basis of the traits being considered in the context of habitats. Although behavioral ecology is a subject that is rapidly gaining in momentum, the genetics of most of the organisms studied by sociobiologists is usually rudimentary, or at the best, fragmentary. Ecobehavioral genetic studies on *Drosophila* mating may, therefore, have a role in this complex and diffuse field. It would be an advance upon the main thrust of sociobiology, which consists of observing what occurs in nature and then speculating upon underlying mechanisms. As Kempthorne (1981) points out, experimental sociobiology utilizing some of the standard methods of quantitative genetics has hardly begun. A major problem is that the classical methods of quantitative inheritance do not readily permit the assessment of phenotypic variation in its own right at the *population* level. Slatkin (1981), however, has considered the partitioning of variance among lines in a subdivided population and developed a concept of *populational heritability* for the examination of the establishment and maintenance of between-population genetic differences. The populational heritability is analogous to the heritability of individual characteristics. The isofemale strain approach (Section 4.4) is a population level procedure that should permit at least some tentative steps, especially if some of the traits studied by sociobiologists do have substantial additive genetic variation.

7.4 Larval behavior

Although the behavior of adult *Drosophila* has been the subject of extensive investigations, far less is known about larval behavior despite its certain importance in the development of an organism. Sewell, Burnet, and Connolly (1975) found that larvae of *D. melanogaster* feed

Table 7.4. *Means and ranges of isofemale strains of* Drosophila
melanogaster *and* D. simulans *for number of larvae out of 10*
choosing agar containing 6% ethanol in preference to agar
containing no ethanol after 15 min on a petri dish

	Latitude (°S)	D. melanogaster		D. simulans	
		Mean	Range	Mean	Range
Melbourne	27.5	7.8	8.8 − 7.0 = 1.8	5.5	6.5 − 5.0 = 1.5
Chateau Tahbilk	37	7.5	8.8 − 6.5 = 2.3	5.3	6.2 − 4.6 = 1.6
Townsville	20	6.4	8.8 − 3.6 = 5.2	5.8	6.8 − 5.0 = 1.8

Note:
See Figure 4.4 for techniques.
Source: Data from Parsons (1978b).

continuously during their period of development and the rate of feeding
activity, measured as the number of cephalopharyngeal retractions per
minute, varies with the physiological age of the larva. Feeding rate
responded rapidly to directional selection giving nonoverlapping populations with fast and slow feeding larval strains with realized heritabilities between 0.11 and 0.21 in different selection lines. Crosses between selection lines show significant dominance for fast feeding rate,
which they suggest is adaptive in times of food shortage, when a high
feeding rate would be favored. Locomotor behavior was studied in the
selection lines, but little association was found between locomotor and
feeding behavior, indicating that these behaviors are controlled independently at the genetic level.

The larval stage of the *Drosophila* life cycle is the stage of maximum
resource utilization. It is known from release–recapture experiments
in a winery that adult *D. melanogaster* are attracted to a fermentation
tank during wine making and its sibling species *D. simulans* is not
(McKenzie, 1974). This result is predictable given that *D. melanogaster*
utilizes ethanol as a resource up to high concentrations (Section 3.3).
That is, if the two species are competing in the same medium, the larvae
may occupy different microniches as demonstrated in the field (Section
3.4). It is not, however, as simple as this. Although isofemale strains
from two southern Australian populations of *D. melanogaster* all show
a uniformly high preference for ethanol, only some from Townsville
in the tropics show such high preferences and others show little or no
preference (Table 7.4). This explains the lower mean for the Townsville
population, as well as its extremely high range compared with the two
southern populations. No such interpopulation heterogeneity was
found in *D. simulans* (Table 7.4). As *D. melanogaster* spread south,

there was presumably a premium on selection for ethanol resource exploitation, which can be regarded as a process of selection among isofemale strains. This is demonstrated here by assessment of a behavioral response to a chemically defined resource. The approach to the study of larval resource utilization with isofemale strain comparisons would appear to have considerable potential, since many possible metabolites can be tested. However, there is some need for caution, since Table 4.3 shows that isofemale heritabilities of larval response to ethanol are relatively low.

Begg and Hogben (1946) showed that acetic acid, ethyl acetate, and DL-lactic acid are attractants for *D. melanogaster* adults, and Fuyama (1976) found some interpopulation differences showing the effects of natural selection within this species. Parsons (1979) found that larvae of both *D. melanogaster* and *D. simulans* are attracted to these three compounds with small differences between species and populations compared with ethanol. Not unexpectedly, the differences between these two species of subgenus *Sophophora* and the cosmopolitan species *D. immigrans* of subgenus *Drosophila* are much greater, given that the former two species almost exclusively use fruit resources and the latter fruit and vegetable resources (Section 3.4).

In Section 3.4, the use of ethanol and derivative products acetaldehyde and acetic acid as resources and stresses were discussed in three common sympatric species of the Melbourne area, *D. melanogaster*, *D. simulans*, and *D. immigrans*, and the Adh^{n2} mutant of *D. melanogaster* with predictable results. Considering larvae, it would be predicted that attraction should parallel resource utilization in consequence of adaptation to habitats in nature. This was found to be true for ethanol (Table 7.5). In particular, the Adh^{n2} strain is hardly attracted to ethanol (Figure 7.8), which is to be expected from its alcohol dehydrogenase biochemical phenotype. Hence, at both the interspecific and intraspecific levels, there is an association of biochemical and behavioral phenotypes.

Acetic acid is an attractant for larvae of the three species and the Adh^{n2} mutant. The *D. melanogaster* population has the highest threshold between attraction and avoidance, for which the other two species and the Adh^{n2} strain are clustered. Hence, the expected relationship between biochemical and behavioral phenotypes is less clear for acetic acid than for ethanol. Finally, as expected from the resource-utilization results (Table 3.3), acetaldehyde attracts larvae to a low threshold and the curves are clustered (Table 7.5).

The concept found for ethanol whereby the biochemical phenotype is related to the behavioral phenotype breaks down at low acetic acid concentrations (Figure 7.8), since attraction occurs at concentrations down to one-thousandth of that giving an ethanol response (Parsons

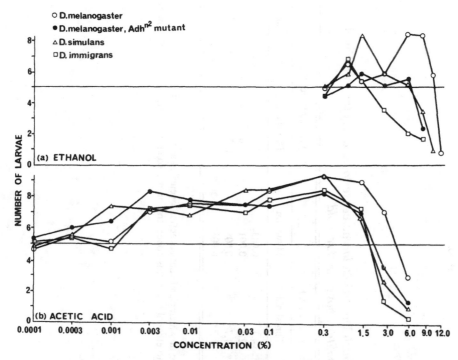

Figure 7.8. Mean number of *Drosophila* larvae out of 10 that choose (a) ethanol or (b) acetic acid for the three species *D. melanogaster*, *D. simulans*, and *D. immigrans* and the Adh^{n2} mutant of *D. melanogaster*. The straight line at five larvae is the threshold between attraction and avoidance of the various ethanol or acetic acid concentrations. (Source: Parsons and Spence, 1981a.)

and Spence, 1981a). These are levels where acetic acid resource utilization is not detectable (Table 3.3). This suggests a more positive role for acetic acid as an attractant for locating food than ethanol. Parallel results were obtained by Fuyama (1976) for adults from an Okinawan strain of *D. melanogaster*, which were attracted to acetic acid in an olfactometer under standard conditions at concentrations of one-thousandth or less of those attracting flies to ethanol. This suggests that acetic acid is both an attractant and a resource for larvae and adults in the wild, whereas ethanol is mainly a resource. Experiments are needed to investigate the distance over which the two metabolites are detectable to verify this interpretation. For both metabolites, however, minimum attraction occurs to lower concentrations for adults than larvae, which is predictable given the greater mobility of adults compared with larvae. Even so, there are probably no chemical substances yet known showing so potent an attractiveness as the totality of naturally

Table 7.5. *Comparison of threshold concentrations at which* Drosophila *larvae cease to be attracted to ethanol, acetic acid, and acetaldehyde and so avoid these metabolites above the threshold*

Drosophila sp.	Ethanol (%)		Acetic acid (%)		Acetaldehyde (%)	
	Threshold	Minimum	Threshold	Minimum	Threshold	Minimum
melanogaster	13.0	1.0	4.1	0.003	1.8	0.1
Adh[n2] mutant	6.0	—[a]	2.1	0.001	1.6	0.1
simulans	6.0	1.0	1.7	0.001	0.8	0.1
immigrans	4.4	1.0	1.7	0.003	1.4	0.1

Note:
See Figure 4.4 for technique. The approximate minimum concentration at which significant attraction occurs is also provided.
[a] Although there is clear avoidance above 6% ethanol, the responses to ethanol below this concentration are so close to the threshold that no real minimum can be established.
Source: Summarized from Parsons and Spence (1981a,b).

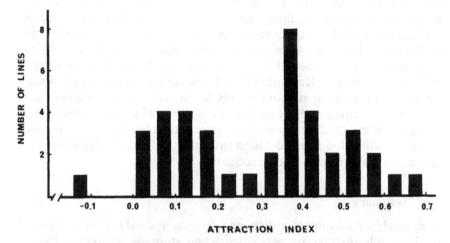

Figure 7.9. Distribution of attraction indices of 40 homozygous second chromosomes extracted from a natural population of *Drosophila melanogaster* adults when tested for olfactory responses to 0.25% ethylacetate in a glass olfactometer. (Source: Fuyama, 1978.)

occurring attractants in fermenting fruits. The high effectiveness of mixtures of compounds, including ethanol and acetic acid, has in fact been shown by earlier workers in studies on adults (Hutner, Kaplan, and Enzman, 1937; Reed, 1938; West, 1961).

From the point of view of genetic studies, therefore, acetic acid appears to offer two components: as a resource and as an attractant. Although Fuyama (1976) did not find strain differences among adults for olfactory response to acetic acid, it would be of interest to see how the components are associated in natural populations both intraspecifically and interspecifically; a one-generation approach at the intraspecific level would utilize isofemale strains (Parsons, 1980a). That reasonably sophisticated genetic analyses may be possible in *D. melanogaster* is shown by Fuyama (1978), who found heterogeneity in natural populations for adult olfactory responses to ethyl acetate controlled by genes on the right arm of the second chromosome (Figure 7.9).

Acetic acid is almost certainly not an odorant with very specific chemical properties evoking a response (Ehrman and Parsons, 1981a). It is more likely to be one of several *resource recognition compounds*, which are normal fermentation products. In this sense, there appears to be an analogy with the recognition compounds suggested by Ehrman and Probber (1978) to provide a characteristic bouquet to males, which provide cues for female mating choices. If acetic acid is a resource recognition compound, then it would presumably need to be present for much of the duration of availability of a resource. The lower volatility of acetic acid (boiling point, 117.9°C) compared with ethanol (boiling point, 78.5°C) argues for the potential importance of acetic acid in this role. This would be of greater significance for isolated quantities of resources than for the rather special grape residue situation analyzed by McKenzie and McKechnie (1979). More generally, any geographical differences found among populations for attractants in isolation may relate to the nature and concentration of the total array of attractants in the habitats of flies. Indeed, the relation between attraction and resource-utilization needs detailed investigation, which should be extended to many fermentation products.

7.5 Summary

Phototaxis, or movement with respect to light, is used to illustrate some of the emphases that are necessary when studying behavioral traits. For accurate genetic analysis, precisely defined environments are mandatory. There may be real difficulties in the assessment of the significance of a laboratory-analyzed behavior in nature. In the case of phototaxis, the problem is made more difficult because of the various ways in which the trait can be measured. Other light-dependent behaviors include oviposition and mating behavior. Oviposition is light dependent in a way that depends upon habitat at the latitudinal level, whereby oviposition occurs mainly in darkness in flies from equatorial regions but the proportion of eggs laid in the light increases away from the equator.

Based upon field observations, flies are not randomly distributed in a forest habitat but tend to select microhabitats meeting their specific requirements, depending upon temperature/humidity/light intensity relationships. Resource utilization occurs within relatively narrow temperature ranges, depending upon the species. Few studies of variability at the intrapopulation level have been carried out. However, isofemale strain studies for a range of developmental temperatures in *D. melanogaster* and *D. simulans* revealed substantial temperature, strain, and temperature × strain effects for a number of traits including percent of matings and development time, which may increase the capacity of a population for adaptation in a variable habitat.

Although mating in *Drosophila* has been extensively studied, here it is considered from three points of view: (1) ecology, (2) ethology, and (3) evolutionary genetics. At the ecological level, intraspecific studies within and among populations under a range of environmental conditions are needed. The ethological evidence stresses the importance of sexual selection among males rather than among females, especially in lek behavior situations. From evolutionary genetics studies in the laboratory, there is good evidence for the importance of male reproductive success. Lek species, in particular, provide indications of the physical features of the environment needed for mating, and, hence, the possibility of the beginning of a link of ecology, ethology, and evolutionary genetics, that is, an ecobehavioral genetics of mating. More generally, there is a great need to study mating under an array of environments, especially those simulating the wild.

Larval behavior of *Drosophila* has been neglected compared with studies on adult behavior, despite the importance of the larval stage in the development of an organism. For ethanol, adult utilization parallels larval attraction, but for acetic acid the relationship between biochemical and behavioral phenotypes becomes more obscure, because acetic acid is likely to be a resource recognition compound as well as a resource.

8

Habitat selection

It is a fact within the experience of most persons, that the various species of animals are not uniformly distributed over the surface of the country. [Wallace, 1876, Part I:3]

8.1 Habitat choice models

Although this book commenced mainly with ecological considerations, the last chapter was where the first major considerations of an eco-behavioral approach were presented. In any case, the development of a coherent evolutionary biology of an organism in its habitat depends upon a unification of genetics, ecology, and the study of behavior. For many organisms, we lack the necessary background information for attempts at unifying the three entities. There is also the difficulty of demonstrating the connection between the laboratory and the field. This requires the study of associations among species and genetic variants within species at the population level, with habitats chosen and resources utilized in nature. In the last few years, this is beginning to be achieved especially in *Drosophila*, so that at the intraspecific level, ecological and behavioral genetics are becoming merged into ecobehavioral genetics. Emphasis on this area of genetics should increase, given the rapid development of the hybrid fields of behavioral ecology and behavior genetics, although ecological genetics has developed over a longer period of time. Ecobehavioral genetic studies of traits involved in habitat selection are important for our understanding of evolutionary processes in natural populations.

Much of the history of evolutionary genetics is based upon models where fitnesses are assigned to individual genotypes and are constant. Under this assumption, genetic variation in a population leading to a polymorphism is maintained under the model, whereby the heterozygote (Aa) at a locus has greater fitness than the corresponding homozygotes (AA,aa) so that $Aa > AA,aa$, that is, there is heterosis or overdominance. However, there are remarkably few well-documented examples, although the human hemoglobins provide an example, since in malarial regions the sickle-cell heterozygotes ($Hb^A Hb^S$) have a higher fitness than the common homozygotes ($Hb^A Hb^A$), which exceed the fitness of the sickle-cell homozygotes ($Hb^S Hb^S$), that is, $Hb^A Hb^S > Hb^A Hb^A > Hb^S Hb^S$. Note, however, that this relationship is environment dependent, since in nonmalarial regions the heterozygote advantage disappears, and the fitness ordering becomes $Hb^A Hb^A > Hb^A Hb^S > Hb^S Hb^S$. Accordingly, it is more logical to regard fitness as a *phenotype*, that is, the realized manifestation of a trait, whereby

$$\text{phenotype} = \text{genotype} + \text{environment}$$

which means that the selective value of an individual is determined by its genotype and the environment in which it lives. If the fitnesses vary spatially, that is, across environments, then the conditions for the maintenance for stable polymorphisms are relaxed compared with heterosis/overdominance models. An early model (Levene, 1953) assumed that zygotes are distributed randomly to habitats. Depending upon the environment of the habitat, one genotype or another will survive better. Those genotypes favored in one habitat may be at a disadvantage in another. Theoretically, this is a way of maintaining a high level of genetic variability with much lower overall fitness differences than the heterosis model. This is of considerable evolutionary significance, since it has been argued that there is a limit to the amount of variability a population can maintain under the overdominance model, because of the associated *genetic load* due to the fitness differences between heterozygotes and homozygotes [see Wallace (1970, 1981) for a good discussion of the arguments involved].

Habitat choice models form a modification of the Levene model. Under the Levene model, genotypes are randomly distributed and fitnesses depend upon the environment occupied. Habitat choice models assume that genotypes avoid areas of the environment in which they are least fit and concentrate in areas where their fitnesses are greatest. In terms of Wallace's (1981) dichotomy of hard versus soft selection, perusal of Figure 2.1 shows that habitat choice involves mainly soft selection. Both the Levene and the habitat-choice models, therefore, obviate the need for uniform heterosis, and both postulate a high degree of substructure to the population. In the case of the Levene model, the structure arises from selection, whereas with habitat choice it comes about through behavioral preference as well as selection, that is, the habitat preference model is ecobehavioral, whereas the Levene model is ecological. Both models predict that environmental diversity leads to genetic variation with a minimum of genetic load. Indeed, multiallelic polymorphisms that are difficult to interpret under constant fitness models become readily interpretable. Habitat choice would appear to imply in some circumstances that individuals vary physiologically in some way corresponding to the environmental variability of a habitat, that is, they may be quite sensitive to their external and internal environments and adjust to these variations behaviorally. Templeton and Rothman (1981) consider that organisms displaying much habitat selection should have poor physiological long-term, but effective physiological short-term, avoidance mechanisms.

8.2   Habitat selection – examples

Powell and Taylor (1979) studied 12 mutant strains of *Drosophila pseudoobscura* on 10 different media at two temperatures in the laboratory. Eggs of two of the strains were placed at the junction of two media, so that upon hatching, larvae made a choice of the preferred medium. It was found that there was a positive correlation between emergence and the medium chosen, and also larvae chose the medium in which they developed fastest. This is important since, as we have already seen, development time to adult is a particularly important fitness component. This means that genetic diversity is positively correlated with environmental diversity, and given a choice, genotypes may tend to congregate in those environments in which they have high fitness.

Using *D. simulans* adults, Jones and Probert (1980) studied matings of white-eyed (*w*) and wild-type strains in three light regimes in three cages: (1) exposed to white light, (2) exposed to red light, and (3) half the cage exposed to white light and the other half to red light, making a heterogeneous environment. After 30 weeks, the gene frequency *w* became 0.01, 0.06, and 0.32 in regimes 1, 2, and 3, respectively. A polymorphism has developed in the heterogeneous environment cage presumably maintained by habitat selection by flies in the two halves of the cage, which was subsequently confirmed by one-generation choice experiments in such cages.

Another example of a behavioral polymorphism arose in a laboratory population of *D. willistoni*, whereby the number of adults in population cages suddenly increased dramatically (de Souza, da Cunha, and dos Santos, 1970). Accompanying this increase in population size, large numbers of larvae began pupating on the floor of the population cage, outside the food cups in which larval development normally occurs. A genetic analysis revealed that the difference between pupation sites was determined by two alleles at a single locus, the allele predisposing larvae to pupate outside the food cup being dominant to its homologue. The effect of this polymorphism is shown in Table 8.1. Cages B and C that were started with one type of fly *or* the other had equilibrium sizes of about 4000 to 4500 flies, and cage E, which was started with recently caught flies, grew to about the same size. However, cages A and D that were started either with a mixture of the two strains or with

Table 8.1. *The dependence of population size in laboratory populations of* Drosophila willistoni *according to genotype for pupating inside the food cups and outside the food cups on the cage floor*

Population	Initiating population	Ultimate equilibrium size
A	750 in + 750 out	7060
B	1500 out	4000
C	1500 in	4280
D	1500 in/out hybrids	7070
E	1500 flies from the wild	4250

Source: Data from de Souza, da Cunha, and dos Santos (1970).

hybrids between them grew to equilibrium sizes more than half again as large.

The advantage accompanying the polymorphism for pupation site lies in the exploitation of an otherwise unoccupied territory within the population cage. If those larvae pupating on the floor of the cage had remained in the food cup, they would presumably have died or caused the death of other larvae. This polymorphism is clearly an example of soft selection, being both frequency and density dependent.

There are a number of *Drosophila* species that normally pupate in the soil and are very difficult to culture unless pupation sites outside the food cups are provided, for example, many of the Hawaiian *Drosophila* species (Carson et al., 1970). Another example is the *coracina* group species of subgenus *Scaptodrosophila*, where on maturity the larvae of some species have a tendency to "skip", that is, they are capable of jumping or springing movements after coiling their bodies into semiloops (Bock and Parsons, 1978), presumably as an adaptation to separate pupation sites. In many of these species, if separate pupation sites are not provided, the form of selection is basically hard, since the species themselves are threatened without such habitat heterogeneity. Even so, certain species, such as *D. (Scaptodrosophila) lativittata*, can be cultured in a single container with difficulty, by providing special pupation sites within the container.

In natural populations if genotypes have different fitnesses in differing habitats, then in an ecologically diverse area genotypic frequencies may vary across habitats. Taylor and Powell (1977) carried out field studies with *D. persimilis* in the Sierra Nevada Mountains of California in an area heterogeneous for physical and biotic factors (moisture levels and vegetation types). Over fairly short distances, gene and inversion frequencies differed from habitat to habitat. After considering several models of selection and migration, they concluded that selection (in the sense of differential viability, survival, or fertility of genotypes) could not by itself explain the results observed. They concluded that the behavior of different genotypes must be different, that is, there is habitat choice based upon the genotype. This was confirmed by release–recapture experiments, whereby flies tended to return to habitats similar to those in which they were captured. That is, flies from ecologically similar areas behaved similarly, irrespective of geographical origin. Hence, flies perceived differences in ecological factors and chose accordingly. However, care is needed in interpreting data of this type, since Endler (1979) has put forward a theoretical model under which a selective interpretation is much more likely, and he suggests that "site fidelity may simply be a result of flies moving more slowly in favorable microhabitats compared to unfavorable ones." Interpretations become very difficult because of a lack of knowledge of the spacing of the breeding sites. As he comments, "we clearly need much more data on population structure and life history parameters." Whatever the interpretation, genotypes may tend to be congregated in habitats where they have higher than average fitness in the previous generation. Over several generations, such behavior may approach optimal habitat choice at the genotypic level. It is, therefore, possible that habitat choice may be important in explaining microgeographical genetic differentiation. Similar microgeographical differentiation has been found in *D. melanogaster* (Stalker, 1976) for karyotypes of flies breeding on grapefruits and oranges, respectively. Clearly one potential advantage of *Drosophila* is that ecobehavioral genetic studies in the laboratory may with care be relatable to the field situation and vice versa.

Evidence for intraspecific habitat selection is not restricted to *Drosophila* (Mayr, 1963; Partridge, 1978). Examples in other species include the pigment polymorphism in the moth *Biston betularia* (Kettlewell, 1973), habitat selection by adults of differing karyotypes of the mosquito *Anopheles arabiensis* of dry versus humid environments (Coluzzi et al., 1979), the differential migration of amylase genotypes of the crustacean *Asellus aquaticus* in a small pond where decaying beech leaves are the dominant food resource in one sector, and decaying willow leaves in another (Christensen, 1977), and a correlation between larval habitat differences (oak treehole vs. beech treehole) and gene-frequency changes at an esterase locus in the mosquito, *Aedes triseriatus* (Saul et al., 1978). As a final example, planktonic larvae of the marine polychaete, *Spinorbus borealis*, settle and metamorphose on several algal species with marked local variations in algal substrate preferences (Doyle, 1976; McKay and Doyle, 1978). There is a tendency of animals to rank a habitat higher on a scale of preferences if the parents came from that habitat.

There is a problem in many of these studies, since the associations of genotypes with habitats are mainly based upon genotypes defined in terms of gene (in particular, electrophoretic) and chromosome polymorphisms. Since these genotypic assessments are not directly relatable to the field situation, interpretations often become difficult. Certainly, genotypes assessed in this way provide estimates of variability levels. However, of greater intrinsic significance would be to take those ecobehavioral traits involved directly in determining the distribution and abundance of organisms as the primary phenotypes for study. Such ecobehavioral phenotypes should relate to those ecobehavioral variables determining which habitats are preferred and would relate directly to the abiotic (temperature, humidity, light intensity) and biotic (resources utilized) features of the environment. There are signs of a start in this regard, since Kekić, Taylor, and Andjelković (1980) have found with capture–mark–recapture experiments that phototaxis appears to play a role in the way *Drosophila subobscura* is distributed in the wild. Indeed, females from neighboring light and dark areas differed in phototaxis as measured in the laboratory, and offspring from flies collected in neighboring light and dark areas were also different for this trait. Although more data are essential, this suggests that there may be genetic differences among flies living only a few meters apart, indicating the possibility that these populations may show a discernible microstructure at the ecobehavioral level.

Quantitative ecobehavioral phenotypes can be analyzed at the population level in species, such as *D. melanogaster*, using isofemale strains as the starting material as one approach. Parallel considerations at the ecological level emerged from a consideration of the adaptive systems of colonizing species discussed earlier in this book. This is reasonable, since a colonization event in itself implies the occupation of a new habitat. Hence, by selecting ecobehavioral phenotypes as the primary study material, it should ultimately become possible to determine relationships with gene and chromosome polymorphisms, even though they may be complex.

8.3 Learning and resource-utilization races

Manning (1967) found that *Drosophila* larvae reared on a medium containing geraniol (a peppermint smell) showed reduced aversion to the odor when adult. He considers this to be a form of habituation. It is reasonable to postulate that over a number of generations, there could be genetic assimilation for such behavior as suggested by Moray and

Connolly (1963). In the wild, if a niche has a resource alteration that can be assimilated genetically, the possibility of rapid evolutionary change occurs for the utilization of different resources. Studies separating learning from habituation are obviously important (Hershberger and Smith, 1967; Manning, 1967). Indeed Quinn, Harris, and Benzer (1974) have shown that *D. melanogaster* can learn a complex discrimination task involving the selective avoidance of an odorant after being shocked in its presence. The two odorants were 3-octanol and 4-methylcyclohexanol, one being coupled with electric shock. Different wild-type strains vary in abilities to display such learning via conditioning (Dudai, 1977). Even so, conditioning, which involves reorientation via training has been difficult to demonstrate reliably in the Diptera, but several reports have now been published in *Drosophila* for various odorants and light stimuli (e.g., Spatz, Emmens, and Reichert, 1974; Quinn and Dudai, 1976) and the blowfly *Phormia regina* (McGuire and Hirsch, 1977). If these results can be related to the avoidance of noxious substances in nature, then they have considerable evolutionary significance. At the theoretical level, the significance of this result is emphasized by Templeton and Rothman (1981), who argue that short-term physiological avoidance mechanisms are important in determining habitats selected. The converse is learning to utilize specific resources in nature, which would lead to ecobehavioral divergence in response to habitats. In this context, the results of Atkinson and Miller (1980) in *D. subobscura* are of interest, since they found no evidence for individual habitat choice. Rather, there was some suggestion of breeding site fidelity, with individuals returning to familiar baits (banana vs. malt). However, this is a one-generation experiment, and more data over several generations are needed to explore the possibility of genetic assimilation.

Memory in the aforementioned conditioning experiments in *Drosophila* appears not to last more than 24 hr. In larvae of *D. melanogaster*, Aceves-Piña and Quinn (1979) found memory in similar experiments that lasted for an even shorter period of time. It would seem that the life span of *Drosophila* may be too short for the development of learning beyond a rudimentary level as a factor in habitat selection. Rather, suitable habitats for insects, such as *Drosophila*, would appear to depend intimately upon appropriate physical conditions as a prerequisite for resource utilization. However, higher in the phylogenetic series where behavioral factors become progressively more important in insulating organisms from the effect of microenvironment, learning should assume greater importance in the selection of habitats. In any case, the heterogeneity of resources of many of the widespread *Drosophila* species may preclude much gene-pool fragmentation of this type. This may be more likely in rarer more specialized species (Section

9.3). Turning to nearctic *Pieris* butterflies, larvae of five species accept a wide range of native and naturalized crucifers under laboratory test situations. The preferences among crucifers tend to be statistical rather than absolute – as in cosmopolitan *Drosophila* species (Atkinson and Shorrocks, 1977). Preferences are not significantly altered by larval development or prior experience with specific food plant species. This means that there is behavioral flexibility in *Pieris* enabling older larvae to exploit alternative resources not available for young larvae (Chew, 1980). This does not exclude the existence of individuals that specialize on one crucifer species, but exposure to certain foodstuffs does not appear to induce such specificity – so demonstrating the unimportance of phenomena, such as habituation and learning, in these resource generalist species.

In the Mediterranean honeybee, which has a longer life span than widespread *Drosophila* species, there is learning for resources (Lindauer, 1975). Various races and species were trained with 17 odors. Naive bees of the different races learned these odors with varying degrees of ease; however, each race learned with the greatest facility the odor of its own native flora. Interestingly, the odor of orange was not learned; however, oranges are from southeastern Asia and have only been cultivated in the Mediterranean for 300 years. Apparently, this is insufficient time to allow western honeybees to become adapted to such food resources. Therefore, learning of resources is strongly race specific in honeybees and so has a genetic component.

Partridge (1981) confined four small mammals, the house mouse, *Mus musculus*, the bank vole, *Clethrionomys glareolus*, the short-tailed vole, *Microtus agrestis*, and the woodmouse, *Apodemus sylvaticus*, to diets of oats or wheat and found that individuals of these four species showed increased preferences for the more familiar diet. She argues that the evolutionary significance of this result concerns an increase in digestive efficiency on current diets. Perhaps enzymes secreted or the nature of the microorganisms in the gut may be of importance, so increasing the rate of food assimilation. Thus, the food in an animal's diet may have a higher nutritional value than the same food when not recently eaten. Therefore, an optimal food choice will vary according to what an animal has recently eaten. If diets do not regularly vary, such differences could lead to resource-utilization races.

A parallel type of observation in *Drosophila melanogaster* is the finding that significantly more larvae went to the arm of a maze containing their own strain (as larvae or biotic residues) as a stimulus, suggesting olfactory discrimination among the larvae, which may be of importance in the selection of habitats (Pruzan and Bush, 1977). Perhaps the yeasts carried by *Drosophila* would repay more study in investigating the possibility of resource-utilization races (see Starmer,

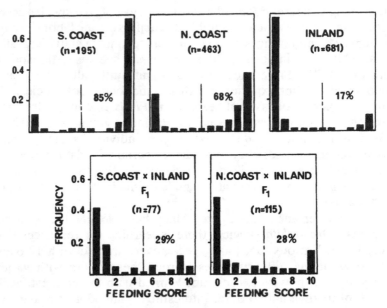

Figure 8.1. Distributions of slug feeding scores in three parental populations of *Thamnophis elegans* (above) and their F₁ hybrids (below). Percents indicate the proportions of naive, newborn snakes that were consistent slug eaters (feeding score = 5 to 10.) (Source: Arnold, 1981a.)

1981). Even so, ecobehavioral genetic studies of learning and adaptation to resources may well be more profitably applied to organisms higher in the phylogenetic series, such as rodents, rather than widespread species of *Drosophila* that are easily cultured in the laboratory and tend to be resource generalists in the wild.

With these comments in mind and turning to newborn garter snakes, *Thamnophis elegans*, Arnold (1981a) studied feeding responses of three parental populations and their F₁ progeny (Figure 8.1). Coastal populations that are sympatric with slugs show the highest incidence of slug-eating snakes (68% to 85%), whereas in an inland allopatric population the incidence is only 17%. The geographical variation in this stable behavior difference is undoubtedly maintained by selection. Crosses between populations indicate at least partial dominance for slug refusal, and reciprocal crosses show similar incidences of slug eating (19% to 32%) providing no indication of maternal effects (Figure 8.1). The phenotypic correlations between responses to 10 different prey odors were partitioned into genetic and environmental parts, by analyzing variation and covariation within and among families (litters of full sibs). Factor analysis of genetic correlations (Arnold, 1981b) is suggestive of one group of genes affecting responses to amphibians (anurans and salamanders), another group affecting responses to both slugs and leeches, and a third affecting responses to the toxic salamander, *Taricha*. These genetic correlations suggest that some genes affect chemoreceptive responses to particular molecules that are shared by certain, often related, prey species. Because of the genetic correlations, selection to recognize or avoid one species of prey would affect predatory reactions to many other species of prey. For example, the genetic correlation between responses to slugs and leeches was extremely high (0.89), and geographical variation in these responses was in agreement. Here then are resource-utilization races under direct genetic control. Even so, the heritabilities of these chemoreceptive responses to prey tend to be in the low to moderate range (20% to 35% as upper estimates). These heritabilities are consistent with larval ethanol preferences in *Drosophila melanogaster* (Table 4.3), which are low compared with those obtained for longevity over relatively stressful concentrations of ethanol. Heritabilities in this range pose difficulties for the genetic analysis of resource races, but with careful design such difficulties should not be insurmountable in laboratory rodents or even *Drosophila*, where tests for learning/habituation could be incorporated in addition.

8.4 Host–plant variability

Whitham (1981) investigated the reproductive success of gall aphids, *Pemphigus betae*, on cottonwood trees, *Populus angustifolia*, at the individual level. He comments that although "ecologists generally try to minimize variation and deal with the mean individual or mean population . . . in the case of plants, evolution may have favored a high level of variation within individuals." He considers that "variation within a plant may be produced by at least four mechanisms, three of which are genetic: (1) induced factors such as pest-induced plant defences, (2) somatic mutations and chimeras, (3) developmental patterns of plant growth, and (4) epigenetic factors."

In early spring when leaf buds first begin to break, the colonizing stem mothers of *P. betae* emerge from overwintering eggs in the fissures of the tree trunk and migrate to the developing leaves. Within 4 or 5 days of bud break, most colonizing stem mothers have permanently settled, and gall formation has been initiated. Within a gall, up to 300 progeny are produced parthenogenetically, so that population growth is largely achieved by asexual reproduction. Since each stem mother produces either a successful gall or dies during gall formation, leaving a scar as evidence of the attempt, a census of successful and aborted galls accurately reflects the colonizing population. By counting the number of progeny within mature galls, the reproductive success of

Table 8.2. *Densities and survival rates of colonizing stem mothers of the gall aphid* Pemphigus betae *on three adjacent branches of the same tree and on five host trees of narrowleaf cottonwood,* Populus angustifolia

	Mean leaf size (cm^2)	Galls/1000 leaves	Percent survival
Adjacent branches			
1	5.0	17.5	36.7
2	8.0	24.2	61.8
3	9.9	477.9	72.1
Trees			
1	4.7	7.7	24.2
2	5.1	0	—
3	6.1	13.8	66.0
4	8.8	105.7	78.5
5	15.9	183.8	87.0

Source: Summarized from Whitham (1981).

surviving stem mothers can be measured. With these data, the patterns and effects of host variation on aphid selectivity during colonization can be examined, together with the resultant reproductive success of individuals. As an example of individual host plant variability are data in Table 8.2, showing the pronounced effects of variation between adjacent branches of the same tree on densities and survival rates of colonizing stem mothers. This clearly demonstrates that the individual host plant is a mosaic of varying susceptibilities to attack. In fact, the variation within a single plant is nearly as great as the variation between trees at the same study site (Table 8.2).

Whitham (1981) suggests that the variation may constitute an adaptive plant response to parasites and herbivores, since it can negatively affect the pest population in various ways:

1. The plant heterogeneity makes the plant less apparent to its pests by increasing the probability that inappropriate settling and feeding decisions will be made so that insect fitness is reduced.
2. The plant heterogeneity increases the level of competitive interactions for the superior plant host resources. Aggressive neglect, floater populations, and other insect behaviors reducing the cumulative effect of the pest on the plant occur as a result.
3. The clumping of parasites and herbivores on the most favorable host resources makes them more obvious, and so they are more vulnerable to their predators.

Colonizing aphids thus respond to individual host plants as if they

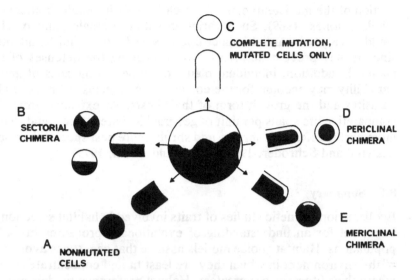

Figure 8.2. This diagram shows that buds derived from plant chimeras can vary greatly in their genetic composition. [Source: Whitham and Slobodchikoff (1981) derived from Neilson-Jones (1969).]

are a mosaic of habitats with varying susceptibilities to successful colonization. Such heterogeneity of the host plants together with predictable responses of the parasites may allow the plant to evolve adaptations so that the effects of pests are minimized in impact.

This example is cited because of the intimate association of successful colonization with habitats; genetic analyses of the colonists would be of great interest. The second reason for citing it is that significant results come from concentrating upon variation among plants within populations – and of greatest importance in this case, variation within individuals. Somewhat parallel data are presented by Edmunds and Alstad (1981) for the colonizing ability of the black pineleaf scale, *Nuculapsis californica*, on the Ponderosa pine, *Pinus ponderosa*.

The high within-plant variability reflects the differing role of somatic mutations in plants and animals (Whitham and Slobodchikoff, 1981). A plant is composed of many repeating modules, each capable of reproduction (Harper, 1977), whereas animals consist of mutually interdependent systems that cannot compete. Consequently, plants are able to take advantage of somatic mutations in ways unavailable to animals. Somatic mutations arising in plants can be inherited by naturally occurring mechanisms of sexual and asexual reproduction. Long life span, large clone size, and the complete regeneration of buds each year may permit an individual plant or clone to evolve. Indeed plants may even develop as mosaics of genetic variation. Figure 8.2 indicates that buds derived from chimeras (derived from mutations only affecting a portion of the meristem) can vary greatly in their genetic composition (Neilson-Jones, 1969). Such variation within a single plant or clone could have a substantial impact upon its herbivores and be adaptive, since it may prevent herbivores from breaking the defenses of host plants. In addition, individual plants or clones as mosaics of genetic variability may account for the correlation of species richness of plant parasites and the growth form of their hosts. As expected from their comparative life spans per unit of geographical area, trees tend to have more pest species than shrubs, and shrubs more pest species than herbs (Lawton and Schroder, 1977; Strong and Levin, 1979).

8.5 Summary

Ecobehavioral genetic studies of traits involved in habitat selection are important for an understanding of evolutionary processes in natural populations. Habitat choice models assume that genotypes avoid areas of the environment in which they are least fit and concentrate in areas where their fitnesses are greatest. Habitat preference models are eco-behavioral and predict that environmental diversity leads to genetic variation with a minimum of genetic load.

There are increasing numbers of examples of habitat selection in the literature in a number of organisms, principally *Drosophila*, but also mosquitoes, crustaceans, and marine polychaetes. In most cases, associations of genotypes with habitats are based upon genotypes defined in terms of gene and chromosome polymorphisms. Interpretations become difficult because these genotypic assessments are not necessarily directly relatable to the field situation. Using ecobehavioral phenotypes directly involved in habitat selection as the primary study material, it should become possible to determine relationships with gene and chromosome polymorphisms, even if complex.

Drosophila can learn a complex discrimination task involving the selective avoidance of an odorant after being shocked in its presence. However, memory appears not to last more than 24 hr, and in larvae for a shorter period of time. The life span of *Drosophila* may be too short for the development of learning beyond a rudimentary level as a factor in habitat selection. In addition, the heterogeneity of the resources of many of the widespread *Drosophila* species may preclude much gene-pool fragmentation of this type. Resource races, based on chemoreceptive responses, occur in garter snakes. Genetic analysis is somewhat difficult, since the heritabilities of chemoreceptive responses tend to be low.

Turning to plants, there is high variability within plants for the reproductive success of gall aphids on cottonwood trees. This reflects the differing role of somatic mutation in plants and in animals, whereby plants may develop as mosaics of genetic variation. Therefore, colonizing aphids respond to individual host trees as if they are a mosaic of habitats with varying susceptibilities to successful colonization. Indeed, the variation within a single tree is often nearly as great as the variation between trees at the same study site.

9

The ecobehavioral phenotype: generalists and specialists

In order to understand the many curious anomalies we meet with
in studying the distribution of animals and plants, and to be able to
explain how it is that some species and genera have been able to
spread widely over the globe, while others are confined to one
hemisphere, to one continent, or a single island, we must make
some inquiry into the different powers of dispersal of animals and
plants, into the nature of the barriers that limit their migrations,
and into the character of the geological or climatal changes which
have favoured or checked such migrations. [Wallace, 1880:70]

9.1 The ecobehavioral phenotype

On constructing the phylogeny of a group of animals, such as ducks
and pigeons, based on behavioral traits, that phylogeny is extremely
similar to an independently assigned one based upon strictly morphological traits. The interpretation of this parallelism is that both sets of
characters are the product of the same genotype, representing a closed
genetic program in the sense of Mayr (1974), who gives other examples
including gulls and storks. An excellent insect example is a survey of
101 species of *Drosophila* (Spieth, 1952). Generally the evolution of
mating behavior parallels the morphological evolution of the group. In
addition, Spieth concludes that the divergence of mating behavior between species occurs first at the physiological and behavioral levels
and visually observable morphological differences arise much later. In
a later study, Brown (1965) found a similar high association of morphological and behavioral traits for the *obscura* group of species. It
is, for example, clear that both behavioral and morphological differences between mutants within species are slight, those between sibling
species are greater, and those between nonsibling species in the same
division are greater still. The same applies to the level of subgenera
that may show major differences in behavior and morphology, and
Spieth (1974) reiterated this general view after incorporating the Hawaiian Drosophilidae.

It can, therefore, be presumed that under suitable conditions be-

150

Table 9.1. *A comparison of the behavioral phenotypes of two subspecies of the deer mouse,* Peromyscus maniculatus

Behavioral phenotype	*P.m. bairdii*	*P.m. gracilis*
Normal habitat	Prairie	Woodland
Thermotactic optimum (°C)	25.8°	29.1°
Removal of sand from tunnel (lb)	5.9	0.1
Clinging response	Low	High
Locomotor response	High	Low
Preference for laboratory environment		
Field	Prefers	
Woodland		Prefers

Note:
See text for references.

havioral differences in courtship within species could become the prime traits differentiating closely related species – and eventually associated morphological differences would become apparent. The more general question of extrapolating to behavior apart from courtship behavior has been more rarely posed, although mutants and evidence of genetic control of other behaviors are well known as indicated for some traits in Chapter 7. Indeed, although many authors regard animal speciation as the development of reproductive isolating mechanisms, an ability to utilize the resources of the environment in such a way as to be protected against niche competitors (Mayr, 1977) must be regarded as no less important. To some extent, the study of such ecobehavioral characters that may lead to autonomy has been neglected, simply because they are often intrinsically more difficult to study. In this regard, *Drosophila* is no exception.

Good evidence for ecobehavioral phenotypes comes from two subspecies of the deer mouse, *Peromyscus maniculatus bairdii*, the prairie deer mouse, which is a strictly field-dwelling form, and *P.m. gracilis*, which is a woodland form (Table 9.1). The two subspecies differ for a number of traits including temperature preference, digging ability, locomotor response, and clinging response in a way that is compatible with the life histories of the two subspecies (Ogilvie and Stinson, 1966; King, 1967). Various differences in developmental rate assessed by morphological and physiological traits occur between the two subspecies, suggesting the likelihood of developmental and, hence, genetic differences in the morphology and biochemistry of their central nervous systems. Here then we have an association of morphological, physiological, and behavioral traits that must be surely related to habitats occupied – and correspond to artificial laboratory habitat selection studies (Wecker, 1964). In addition, Wecker showed that the prefer-

Table 9.2. *Character combinations found in two types of black mustard,* Brassica nigra, *in western Asia*[a]

Plant part	Race occurring in	
	Open fields	Cultivated fields of yellow mustard
Leaf rosettes in young plant	Well developed	Absent
Branching	Few branches in upper part of plant	Numerous branches from base
Stem	Thick and tall	Slender and short
Lower leaves	Long with 5–7 pairs of lobes	Short with 2 or 3 pairs of lobes
Pubescence	Abundant and rough	Sparse and soft, or lacking
Time of flowering	Late	Early
Inflorescence	Compact	Open
Flowers	Small	Large
Pods	Short and narrow, lying close to stem	Long and broad, standing out from stem
Dehiscence of pods	Pods opening easily, seeds falling out spontaneously	Pods opening with difficulty under pressure, seeds not shedding spontaneously
Seeds	Small	Large

Source: Data of Sinskaja tabulated by Grant (1963).

ence can be reinforced by early experience of the preferred habitat. Ultimately, therefore, so long as the environment remains stable, the population as a whole eventually becomes adjusted to the habitat that it is best able to exploit and shows the corresponding ecobehavioral phenotype at the interpopulational level. The integrity of the components of this phenotype has yet to be investigated at the intrapopulational level, which would need biometric analyses of various segregating generations (Falconer, 1960; Mather and Jinks, 1971).

In plants, the black mustard, *Brassica nigra*, is a coarse weed in open fields in various parts of the world. In western Asia, where the yellow mustard, *Brassica campestris*, is cultivated as a crop plant, the black mustard is represented by another more slender form specialized to survive as a weed in the yellow mustard fields. Populations of black mustard growing with yellow mustard differ from the common weedy form of black mustard found in open fields and waste ground in 11 separate traits (Table 9.2). Collectively, the different traits – mode of branching, stem size, flower size, shape and dehiscence of seed pods, seed size, and time of flowering – work together to help the specialized black mustard race grow and ripen simultaneously with the yellow mustard and to produce seeds, which are threshed and dispersed with those of the crop plant whose field it inhabits [data of Sinskaja summarized in Grant (1963)]. Adaptation as a successful weed in yellow mustard fields is, therefore, based upon a combination of traits enabling precise matching of this habitat by comparison with the open field race and so represents a plant example paralleling the *Peromyscus* subspecies.

The *Peromyscus* work points to a need to study wild house mouse (*Mus musculus*) populations in more detail, since this is an organism upon which genetic studies can more readily be done because the genome is well known. There is certainly variation between inbred strains for traits similar to those discussed in *Peromyscus*, for example, nest building (Lynch and Hegmann, 1972) and temperature preferences (Parsons, 1974b). Using inbred strains and hybrids, associations among traits of ecobehavioral significance have been found (Parsons, 1974b) but additional biometric analyses are necessary.

Stabilizing selection should occur under natural conditions to keep traits important in habitat selection within relatively narrow limits, since animals showing ecobehavioral characteristics away from the norm will be less likely to breed with others. Furthermore, animals occupying the most suitable habitat in a heterogeneous environment have less need to utilize the physiological and behavioral adaptations an animal possesses to combat imperfect environments. Another advantage of occupying a suitable habitat is that the chances of interbreeding by individuals with similar phenotypes are greatly enhanced, thereby ensuring their continuity, so that reproductive isolation could well begin to evolve as discussed later in this chapter.

However, many species, in particular, colonists, are not likely to be as intimately adapted to their habitats as the *Peromyscus* subspecies appear to be. Although stabilizing selection may be the predominant influence, directional selection may frequently occur in response to changing environments, which may occur on a seasonal basis in *Drosophila*. It is, therefore, less likely that generalist colonizing species, which tend to be *r* selected and so are highly variable in population size, will develop intimate ecobehavioral adaptations to specific habitats, as compared with species showing little colonizing ability. The high variability of population size would in itself preclude such an intimate association with habitats.

For example, the crop field habitats occupied by house mice, *Mus musculus*, on a small Maryland farm are characterized by temporal heterogeneity and instability. The demographic patterns of the house mouse population in such habitats fall at the *r* end of the *r*–*K* continuum (Stickel, 1979), with high reproduction and low survival, including rapid

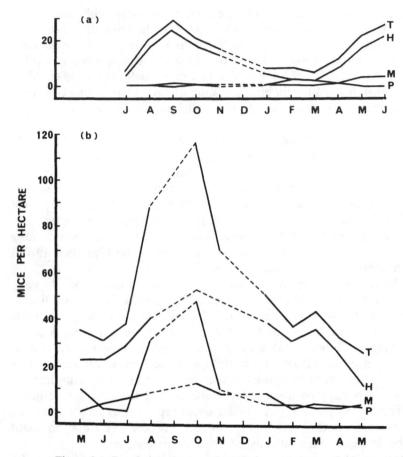

Figure 9.1. Population changes of small mammals in two fields (a and b) in a corn–wheat–hay rotation on a small Maryland farm. The animals were obtained by monthly live trapping over a period in 1949–50. T, Total; H, house mice; M, meadow vole; P, white-footed mouse. (Source: Stickel, 1979.)

growth to maturity, production of several litters in a season, repro-duction by young of the year, and high turnover rates. Population changes are given in Figure 9.1 for the house mouse *Mus musculus*, the meadow vole *Microtus pennsylvanicus pennsylvanicus*, and the white-footed mouse, *Peromyscus leucopus noveboracensis*, showing the extreme variability of the house mouse population especially as compared with *Peromyscus*. In other words, in the course of evolution the members of a species evolve those characteristics that maximize the number of their descendents in their habitats in both time and space, and this may involve varying population size fluctuations among spe-cies (Southwood, 1977; Taylor and Taylor, 1977).

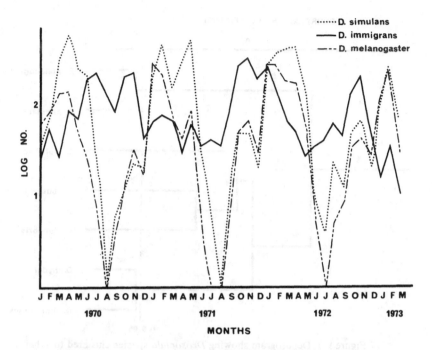

Figure 9.2. Log of the numbers of the three species *Drosophila melanogaster*, *Drosophila simulans*, and *Drosophila immigrans* collected each month in the Melbourne population from January 1970 to March 1973. (Source: Replotted from McKenzie and Parsons, 1974a.)

The comparative population sizes of three sympatric *Drosophila* species of the Melbourne area (Section 3.4) is of interest in this regard. For population size variations on a monthly basis, *D. melanogaster* and *D. simulans* are highly variable since very few flies can be collected in winter months and large numbers during the summer peaks (Figure 9.2). By comparison, *D. immigrans* varies far less, and peak population sizes show much less repeatability on a seasonal basis than the sibling species. Although all three species are cosmopolitan, the sibling species, *D. melanogaster* and *D. simulans*, are obviously closer to the *r* end of the *r–K* continuum than is *D. immigrans*.

In this connection, Atkinson (1979) compared the reproductive systems of seven domestic species of *Drosophila* measured by ovariole number, eggs per ovary, thorax length, adult survival, days to eclosion, relative clutch volume, egg volume, and biomass ratio. A dendrogram (Figure 9.3) shows that the fruit-breeding species *D. melanogaster* and *D. simulans* are clustered together and are distant from the pair of resource generalist species, *D. hydei* and *D immigrans*. Considering all species, Atkinson concludes that species that breed in similar breed-

CORRELATION COEFFICIENT

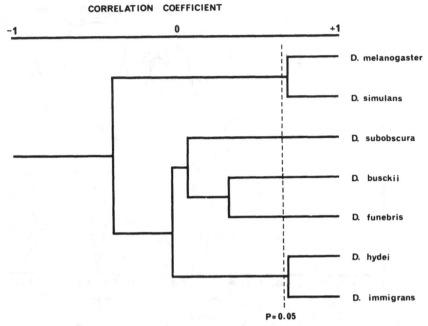

Figure 9.3. Dendrogram showing *Drosophila* species clustered together according to reproductive characters listed in the text. (Source: Atkinson, 1979.)

ing sites tend to have similar reproductive systems. The results are consistent with the conclusion of Kambysellis and Heed (1971) that the larval niches of *Drosophila* must be a major factor in establishing the diversity of female reproductive systems and the suite of characters show some inconsistencies with the *r–K* continuum predictions. Interestingly (Table 9.3), based upon multiple-choice experiments with pairs of strains under test, *D. immigrans* shows reproductive isolation among populations from different geographical localities and a lesser but substantial level among isofemale strains within localities (Ehrman and Parsons, 1981b), whereas *D. melanogaster* apparently shows little evidence for reproductive isolation among populations. Does this mean that *D. immigrans* is closer to evolving discrete ecobehavioral phenotypes among populations than is *D. melanogaster*? Certainly both among and within populations *D. immigrans* has a potentially fragmented gene pool, whether or not there are ecobehavioral correlates.

9.2 Generalists and specialists

A *generalist* is a phenotype whose fitness in one habitat precisely equals its fitness in another habitat. Clearly this is an extreme situation,

Table 9.3. *Comparison of sexual isolation among geographical strains and among isofemale strains of Drosophila immigrans using joint isolation indices (Malogolowkin-Cohen, Simmons, and Levene, 1965), which range between 0 and 1 when there is sexual isolation (and less than 0 when unlikes prefer each other)*

	N	No. (proportion) of positive isolation indices	No. (proportion) of significant[a] positive isolation indices	Maximum isolation index	No. (proportion) of negative isolation indices	No. (proportion) of significant[a] negative isolation indices	Minimum isolation index
Among geographical strains	21	20 (0.95)	15 (0.71)	0.58	1 (0.05)	0 (0)	−0.04
Among isofemale strains	28	22 (0.79)	12 (0.43)	0.34	6 (0.21)	2 (0.07)	−0.24

Note:
An isolation index of 0 indicates random mating.
[a] $P < 0.05$.
Source: Data from Ehrman and Parsons (1981b).

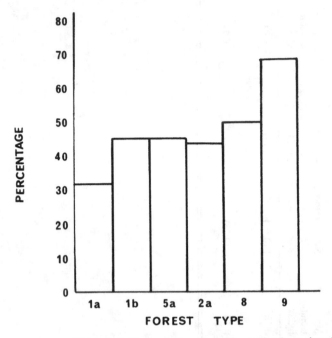

Figure 9.4. Percent of northern Queensland species also occurring from southern Queensland to Tasmania, according to northern Queensland rain forest type (Tracey and Webb, 1975): 1a, complex mesophyll vine forests (lowlands and foothills, altitude <400 m); 1b, complex mesophyll vine forests (uplands, altitude 400 to 800 m); 5a, complex notophyll vine forests (highlands, altitude 800 to 1600 m); 2a, mesophyll vine forests (lowlands and foothills); 8, simple notophyll vine forests (uplands and highlands); 9, simple microphyll vine–fern forests (highlands). (Source: Parsons, 1981a.)

which contrasts with the *specialist* whose fitness is highest in one particular habitat. Correspondingly, the generalist exhibits *fine-grained* behavior, which is the acceptance of resources in the proportions in which they exist. In contrast, the *specialist* exhibits *coarse-grained* behavior, where an organism selects resources in proportions differing from those encountered in the environment (Levins and MacArthur, 1966; Rosenzweig, 1981).

In *Drosophila*, increasing breeding site specificity occurs as species tend toward specialism. In the rain forests of the humid tropics of northern Queensland, only 32% of the species of the low-altitude floristically complex mesophyll vine forests occur further south, compared with 67% from the depauperate simple microphyll vine–fern forests of highland habitats, with intermediate forest types falling between (Figure 9.4). Predictably too, those species common to Victoria and Queensland are very widespread in Victoria and tend to dominate in depauperate communities. Species adapted to the floristically richest tropical forests are, therefore, most unlikely to be found elsewhere, which concurs with the likely sensitivity of such species to environmental extremes (Parsons, 1981a). It is likely that only species such as *D. inornata* have sufficient resistance to environmental extremes (Figure 3.3) to migrate over long distances, as must occur between isolated pockets of suitable vegetation in outlying regions, such as those to the west of Melbourne, Victoria, so that species of depauperate faunas should have the physiological capability to be more migratory than those from richer faunas. Both in the outlying southern cold-stressed habitats and the depauperate habitats of Queensland, which are heat stressed at low and cold stressed at high altitudes, *Drosophila* species appear to fall into some of the more extreme tramp categories of Diamond (1975). Such species are not especially well adapted to any one particular habitat but instead are moderately adapted to a wide range of conditions. In addition, these northern species can usually be cultured in the laboratory, which follows from their "generalist" life characteristics. There is a major need for comprehensive physical tolerance and resource-utilization studies on such species compared with those more restricted in distribution, using groups of closely related species as discussed elsewhere in this book (Section 3.2).

In terms of the flora upon which *Drosophila* depends, Nix (1981) has classified the thermal adaptations of the Australian flora into three broad groups: (a) *megatherms* with thermal optimum around 28°C; (b) *mesotherms* with thermal optimum in the 18° to 22°C range; and (c) *microtherms* with thermal optimum in the 10° to 12°C range. The depauperate *Drosophila* faunas of heat- and cold-stressed habitats appear to fall approximately into the megatherm and microtherm groups, respectively, whereas habitats of highest *Drosophila* diversities fall into the mesotherm group. Field collection data (Section 7.2) are consistent with this pattern.

Another feature of less specialized, and potentially colonizing, species of the extreme habitat categories is that they may be present in a given habitat over a longer period of time seasonally than are more specialized species because of probable breeding site availability restrictions for the latter group (Figure 9.5). This could be the situation for *D. inornata* compared with strict rain forest species. Whether considering climate, soil type, vegetation, or resources, the Australian *Drosophila* fauna shows decreased specialization associated with decreased resource diversity in harsh climates (Parsons, 1981a). Taking temperature in particular, MacArthur (1975) generalized the inverse relationship between environmental fluctuation and species diversity (see also Southwood, 1977). In Wallace's (1981) terminology, it can be said that selection becomes harder with increased temperature varia-

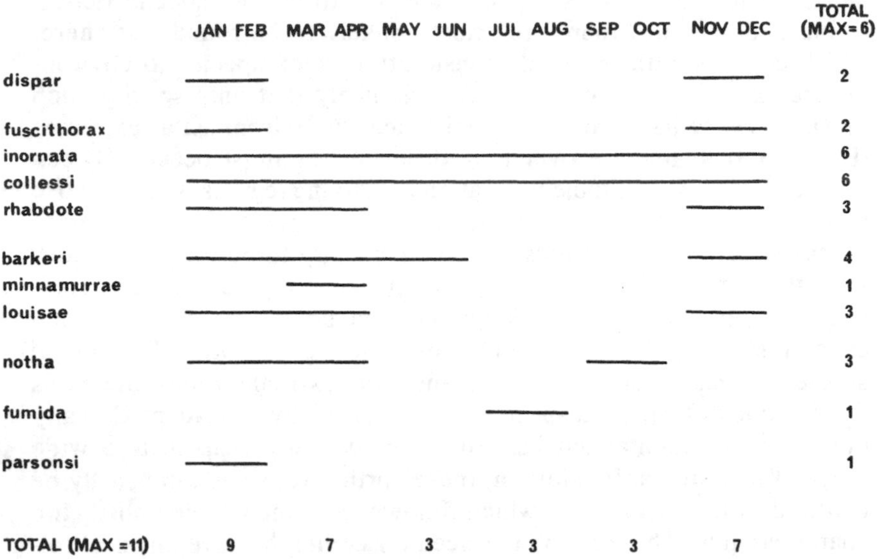

	JAN FEB MAR APR MAY JUN JUL AUG SEP OCT NOV DEC	TOTAL (MAX=6)
dispar		2
fuscithorax		2
inornata		6
collessi		6
rhabdote		3
barkeri		4
minnamurrae		1
louisae		3
notha		3
fumida		1
parsonsi		1

TOTAL (MAX =11)   9   7   3   3   3   7

Figure 9.5. The incidence of 11 species collected by sweeping on a seasonal basis in Kinglake National Park near Melbourne, Victoria. (Source: Parsons, 1981a.)

tion. In Australia, the range is from monospecific populations of *Drosophila* in relatively harsh islands of vegetation in western Victoria and upland regions of southern New South Wales, to the rich complex mesophyll vine forests of northern Queensland, where in some consistently wet habitats in the vicinity of fungi (e.g., Palmerston National Park) there is a tendency in subgenus *Hirtodrosophila* (Table 1.2) toward the close packing of species predicted for tropical biota (MacArthur, 1971). In these consistently wet habitats are the richest and most diverse *Drosophila* faunas so far found in Australia. This agrees with Whittaker's (1977) suggestion that a partial explanation of the innumerable species of tropical regions resides in the accumulation of many plant–fungus–insect interdependences, permitting the survival of many species; these would be conditions of relatively soft selection. It could well be that habitats favoring such species packing are quite rare in northern Queensland rain forests, requiring near-optimality for all the variables of climate, soil type, vegetation, and resources. However, further to the north in New Guinea at appropriate altitudes, there could be much more species packing especially in this subgenus, for which species tend to be more localized in Australia, than for the other subgenera. This would merely be an expression of the observation made on other communities that the average geographical range of species comprising a community declines as diversity increases (Rosenzweig, 1975); presumably selection becomes simultaneously softer because of adaptation to more specialized habitats than for widespread species.

Many authors have presented arguments for an evolutionary trend from generalism to specialism, whereby populations evolve toward a more optimal use of resources (e.g., Dethier, 1954; Feeny, 1975). In *Drosophila*, this implies that specialist resources must be available for a long enough time on an annual basis for population continuity, where a lack of resources does not correspond to the times when physical features of the environment are restrictive. If there is a lack of resources for too long a period, then generalism should theoretically be favored. The North American species, *D. falleni*, breeds on seven morphologically, chemically, and ecologically diverse species of fungi of the genus *Amanita* (Jaenike and Selander, 1979). For four structural genes assayed electrophoretically, genetic diversity was apportioned: 80% to 90% within fungi, 8% to 13% between individual fungi of the same species, and only 2% to 5% between species of fungi. With the proviso that electrophoretic loci may not be good indicators of population divergence at the microhabitat level, these data do not support the subdivision of *D. falleni* into resource-utilization (host) races. The fruit bodies of fungi are highly unpredictable in occurrence and unreliable as resources for specialization, so that resource generalism may be favored.

Carson and Ohta (1981) present a situation in the *D. grimshawi* species complex of Hawaii, where generalism is suggested to be the derivative situation. In these picture-winged *Drosophila*, there is a polyphagous population that oviposits and completes development on 10 families of plants on the island of Maui, whereas the Kaui form oviposits on one plant genus only (*Wikstroemia*). The geological age of the islands suggests that the direction of evolution is from specialist to generalist. Laboratory rearing of these flies indicates that this intraspecific shift in breeding site ecology may largely be the result of ovipositional preferences of the females, rather than nutritional requirements of the larvae. Although a fertility analysis of F_1, F_2, and backcross progenies indicates two sets of coadapted gene complexes differentiating different allopatric populations, the ovipositional behavior difference is controlled by as little as a single-gene two-allele system. This suggests that evolutionary trends in either direction could occur with relative ease.

Drosophila species are usually obtained by one collection method only, implying resource-utilization specificity (Table 1.2). Among the exceptions, certain *coracina* group species of the subgenus *Scaptodrosophila* come to both fermented-fruit and mushroom baits, implying a more generalist life history than many species. Some of these species are reasonably common in urban/orchard areas and have presumably spread subsequent to the introduction of fruit trees after European settlement in Australia (Section 3.2). These ecologically versatile species, which may be in the process of colonizing new habitats, are, therefore, of particular evolutionary interest, since they too may be becoming more generalist. Presumably, the colonization of new habitats could involve ecological polymorphisms related to changes in resource utilization as suggested in Section 3.3 when discussing the alcohol dehydrogenase (*Adh*) polymorphism in these species.

In the case of many host plant shifts, it is likely that the physical features of the environment are very similar. One of the demonstrated features of the ecological phenotypes of colonizing species is a degree of physiological toughness (Section 3.2). Considering the Australian *Drosophila* fauna, widespread and potentially colonizing species do not appear to have the ability to invade undisturbed rain forest habitats in which specialist endemic species have apparently occupied all possible *Drosophila* niches (Parsons, 1981a). Several authors (e.g., Watts, 1971; Webb and Tracey, 1972) have recorded that closed forest communities rarely receive invading plant species, since competition for niches is severe. Because of the dependence of *Drosophila* on plants (and associated fungi and microorganisms) as resources, the lack of success of cosmopolitan species in rain forests is expected. From theoretical considerations, Templeton and Rothman (1981; Section 8.1) argue that habitat selection is most favored in organisms deficient in long-term physiological tolerance, as appears to be the situation in undisturbed rain forest habitats, so providing a contrast with colonizing species.

The Australian *Drosophila* fauna, therefore, shows little overlap between rain forests and urban habitats (Parsons, 1981a). However, as forest habitats at an equivalent latitude become depauperate, the overlap tends to increase, presumably because such forests have more generalist resources than do the richest forests. Even so, the species flow is almost entirely one way, from rain forests to disturbed and urban habitats. In northern Queensland, the major resource feature in common between urban habitats and rain forests are fermenting fruits, which should theoretically permit urban colonizations by fruit feeders of rain forests. However, some species that are very successful in rain forests have not invaded urban habitats, whereas in southern Australia, fermented fruits hardly occur in rain forests, so that urban colonizations would not be expected. Concerning the reverse situation of the invasion of rain forests by the cosmopolitans from urban habitats, *D. immigrans* is a possible candidate if it is assumed that it originally evolved as a citrus specialist becoming a domestic species as these fruits were exploited commercially (Atkinson, 1981; Section 3.4).

It will be of interest to follow the future history of certain urban/

orchard communities since they are the most likely of all to be colonized by additional widespread species from other similar habitats. This could alter the composition of communities as noted in elegant "laboratory island" studies (Wallace, 1975), where more species were found in the absence of *D. melanogaster* than in its presence. In other words, such weedy species may lead to a simplification of communities. However, as noted, it appears that such species are unable to penetrate complex natural communities, as shown by the lack of success of *D. melanogaster* in this regard. In this sense, *D. melanogaster* is a species of ecological margins when considering the genus as a whole. Indeed it is the nonspecialized species generally that tend to be found at ecological margins, where moderate adaptation to a wide range of conditions is at a premium.

Williams (1969) reviewed the iguanid genus, *Anolis*, in the West Indies. Those lizards that have recently expanded their range are of open forest/savanna areas, whereas none are of deep shade, rain forest, or montane habitats. In general, the colonists tend to be in typical unstable habitats, such as edges and clearings. They are neither restricted to, nor very closely adapted to, extreme ecologies, that is, they are neither primitive species nor extreme specialists. Indeed, they are versatile species, being physiologically and ecologically tolerant of many conditions. Colonization voyages must involve desiccation stress levels that rain forest/montane species would find intolerable. Like *Drosophila*, a very limited number of the total known species have achieved successful colonization, and the same species tend to be involved in repeat colonizations. In further considerations of these lizards Lister (1976) argues that niche expansion involves a form of fine-grained habitat selection, whereby a compromise morphology is favored that is capable of exploiting a variety of perch sites, thermally diverse conditions, and an optimal range of prey sites. The conclusions are viewed as a direct consequence of territoriality, which implies fine-grained resource utilization by members of the population (as compared with insects) and generates phenotypic convergence. The compromise phenotype so derived does well in several environments but is a specialist in none, so that the evolution of ecologically versatile generalists is favored.

9.3 Specialism and isolation

Mayr (1963) stressed the importance of behavioral events in colonization when he wrote "a shift into a new adaptive zone is, almost without exception, initiated by a change in behavior." When considering niche shifts in New Hebridean birds, Diamond and Marshall (1977) suggested that colonizing populations may make immediate behavioral adjustments of spatial niche parameters (altitude, height, and vertical position) possibly within the lifetime of colonizing individuals. Later and much more slowly, gene frequencies ought to be systematically shifted, because these behavioral adjustments mean that the population is now living in a different environment and exposed to a different range of resources, different competitors, and, hence, different selective forces. These later changes may affect foraging techniques and food preferences (i.e., resources utilized), and it is argued that these may be subject to fairly rigid genetic programming. This appears to be in agreement with the conclusion of Wasserman and Futuyma (1981) that the diets of phytophagous insects are more labile at the behavioral than at the physiological level; this conclusion came from experiments on the southern cowpea weevil, *Callosobruchus maculatus*, where food preferences for two types of beans could be modified by selection, but there were no concomitant changes in the physiological adaptiveness of larvae. Behavioral and genetic studies of colonists that have lived in altered niches for approximately known times are needed [see also Diamond (1973) on New Guinea birds]. Diamond and Marshall (1977) also argue that there may be a weak but certainly not universal tendency for colonists to become specialized as part of an overall taxon cycle, as put forward by Wilson (1961) when writing on the Melanesian ant fauna.

Animal speciation certainly involves the development of reproductive isolating mechanisms. However, ecobehavioral factors involved in adaptation to new habitats are also important, so that an overall ecobehavioral approach is necessary. Indeed, according to the organism, at a simplistic level a continuum may be expected, ranging from organisms where behavioral factors of a sexual kind may be the most important in isolation to those where ecobehavioral factors of a nonsexual kind are the most important. For example, in certain Hawaiian *Drosophila* species, behavioral factors are regarded as so important that the term *ethospecies* has been used (Carson, 1978) to emphasize the view that the primary change in speciation is the building up of a new coadapted complex in the genome for mating behavior. In a later paper, Carson et al. (1982) have argued that speciation in certain Hawaiian lek species operates through a series of founder events having their principal genetic effect initially on the mating behavior system. In these cases, there is little detectable divergence among species at the inversion or electrophoretic polymorphism level, and ecobehavioral factors of a nonsexual kind appear unimportant. In any case, the relatively benign nature of the environment of Hawaiian forests may be conducive to the development of complex behaviors (Section 7.3).

Kilias, Alahiotis, and Pelecanos (1980) experimentally manipulated temperature and relative humidity in several *D. melanogaster* populations having a common ancestral gene pool. They found that isolation of the populations alone did not lead to ethological isolation but isolation combined with the differing ecological regimes did. The ethological isolation so found was associated with oviposition time differences. However, differentiation among populations for other fitness traits and electrophoretic frequencies was low. In contrast, Ehrman (1964) studied sexual isolation between six populations of *D. pseudoobscura* set up in cages at the same time by M. Vetukhiv and derived from the same initial population. Of the populations, two were maintained at each of 16°, 25°, and 27°C. After a period in excess of 4 years, sexual isolation had evolved between all populations but was as pronounced within temperatures as between temperatures. Isolation has arisen in the absence of any selection for isolation and is evidently a by-product of genetic divergence, but ecological differences did not enhance isolation. This conclusion stresses the problem of interpretation in this complex area, since differences in the species chosen, laboratory techniques, and environments chosen may all be involved. In addition in some species, ecological variants may be the obvious ones to study, and in others, behavioral ones.

The variety of resources utilized within certain species suggests that if a newly entered habitat is characterized by some sort of resource difference, ecobehavioral change is possible in response to the resource change. In the two-spotted spider mite, *Tetranychus urticae*, host races rapidly evolved in adaptation to (1) a monoculture of lima beans and (2) a simple plant community consisting of lima beans and a somewhat toxic host, mite-resistant cucumber (Gould, 1979). Since these races were derived from a single population, this shows that host range evolution can be quite a rapid process in response to resource changes. In *Drosophila virilis*, this possibility was demonstrated by Wallace (1978), who adapted a population to life on decomposing urine, so that within a year, the flies became virtually unable to survive on standard *Drosophila* medium, although the heterogeneity of resources normally used by widespread species would preclude such a situation in nature (Section 8.3). In island situations, however, drosophilids have evolved to exploit an environment analogous to decomposing urine in the nephric grooves of land crabs on three separate occasions (Carson, 1974).

In theory, habitat and resource-utilization heterogeneity in natural populations could under certain circumstances lead to discrete races and possibly even species. This has been suggested for two closely related Hawaiian species, *D. mimica* and *D. kambysellisi*, the latter being proposed as a derivative species that may have arisen following the provision of a new rotting plant resource (Richardson, 1974). This explanation is plausible through an altered food preference followed by genetic assimilation. This possibility is likely for insects that mate

Table 9.4. *Comparison of ecological parameters of* Dacus tryoni *and* Rhagoletis cerasi

	D. tryoni	*R. cerasi*
Maximum observed movement	24 km	500 m
Capacity for increase		
i. Fecundity		
Average eggs per female per week	~80	10
Average eggs per infested fruit	~15	1
ii. Speed of development		
Eggs	2–4 days	10 days
Larvae	9 days (min)	20 days (min)
Pupae	9 days (min)	10–11 months (diapause)
iii. Longevity of adult females	Up to 7 months	Several weeks
Generation time	6 weeks (min)	1 year
Parasitism (maximum observed)	<1%	Up to 32%
Host relationships	Polyphagous (160 known hosts)	Monophagous (1 host)

Source: From summary of Bateman (1977).

on the host plant or animal; for example, in the true fruit flies (Tephritidae), where host races of *Rhagoletis* have evolved over a relatively short time in North America to parasitize domesticated European fruits, such as cherries and apples. Although Bush (1975) has stressed the need for an integrated ecobehavioral approach involving aspects of ecology, behavior, and genetics for the understanding of speciation in these circumstances, more detailed data are needed to accurately assess the evolutionary status of these host races (see for example Futuyma and Mayer, 1980; Jaenike, 1981). Even so, this example fulfils the requirement that the resource is predictable from year to year, which has been argued to be a prerequisite for the development of specialist host races (Jaenike and Selander, 1979) and may explain why species having apparent host races exploit trees in the main.

Bateman (1977) compared the life-history characteristics of the colonizing tephritid, *Dacus tryoni* (Section 2.1), with the European cherry fly, *Rhagoletis cerasi*. As noted already (Table 2.3), the life history of *D. tryoni* fits it extremely well to the transient nature of its larval food supply, and the features of its life history show it to be a typical *r*-selected species. By comparison, *R. cerasi* is more of a *K*-selected species with relatively stable populations that are closely synchronized with the yearly fruiting cycle of one or very few species of host fruits in a limited area (Table 9.4). *K*-Selected pests, therefore, tend to have low rates of potential increase and more specialized food resources than *r*-selected pests (Conway, 1981).

Given the predictable resources implied by *K* selection, sympatric speciation could perhaps occur if mating occurs assortatively within niches containing the resource. Dethier (1954), when writing on olfactory conditioning, considers that if there is assortative mating among members of a population similarly conditioned, then such behavior could be regarded as an isolating mechanism. Maynard Smith (1966) has shown theoretically that if there is a positive correlation between mates and niche selection, speciation may ultimately occur by the development of isolation under the control of a genetic polymorphism developed by disruptive selection. [Additional theoretical considerations on the maintenance of polymorphisms in heterogeneous environments appear in Levene (1953), Mather (1955), and Levins (1968).] Thoday and Gibson (1962) have shown that polymorphism can be developed by disruptive selection for sternopleural bristle number in *Drosophila melanogaster*. Divergence is enhanced if there is a breeding system incorporating positive assortative mating. Later results (Thoday, 1964) showed the development of positive assortative mating in disruptive selection lines. This is reasonable, because unselected *Drosophila* populations may show positive assortative mating for sternopleural bristle number (Parsons, 1965). The experimental procedure was to place virgin females into a mating chamber from which mating pairs are removed with an aspirator. Correlation coefficients between mated pairs were between 0.1 and 0.2 and were all significantly greater than zero. Similar results were obtained for abdominal bristle number. Thus, in these experiments, there is a tendency for flies with similar bristle numbers to mate, which is positive assortative mating. Regrettably, however, sternopleural bristle number is not an ecobehavioral phenotype, so it does not present us with the necessary case study for populations exploiting resources in their habitats, even though the disruptive selection results are of extreme intrinsic interest.

Microhabitat selection in the wild would further promote isolation and is likely once a polymorphism is established (Maynard Smith, 1966). Indeed, microhabitat heterogeneity implies an array of optima in a population dependent upon interactions between genotypes and environments. Formally, such phenotypic habitat selection can be regarded as a form of genotype–environment correlation favoring the development of a coarse-grained array of habitats occupied because an animal selects its environment rather than the environment selecting the animal. Learning associated with habitat selection followed by genetic assimilation would accelerate the development of population heterogeneity and could perhaps be a force in the development of species

(Thorpe, 1945). However, more recently the likelihood of such phenomena has been questioned. For example, Futuyma and Mayer (1980) "have been unable to find a single convincing case in which oviposition by a phytophagous insect is determined by the individual's experience as a larva" (see also Section 8.3).

Considering *Drosophila*, associations between mating and habitat selection are obviously most likely for resource-specific species. *D. hibisci*, which is restricted to endemic Australian *Hibiscus* plants, has only been found on the petals and corolla tubes of *Hibiscus* species, where courtship-type behaviors have been observed (Figure 9.6). Flies spend most of the time in the base of the corolla tube around the filaments of the anthers; this is a humid microniche in a climate that is otherwise extreme for *Drosophila*. Extreme habitat selection is necessary in this species because the *Hibiscus* flowers usually remain open for 1 day only, during which time oviposition occurs followed by larval exploitation of the decaying flowers as a resource (Cook, Parsons, and Bock, 1977). Another *Drosophila* example is the Australian endemic species, *D. minimeta*, which has been found to exploit flowers of two introduced species of the family Solanaceae, namely *Solanum mauritianum* and *S. torvum* in eastern Australia (Parsons, 1981a). These flower-breeding species are, therefore, utilizing long-lasting highly specialized resources of large shrubs or small trees, compared with the more transient resources utilized by generalist species, for which the chances of speciation occurring by habitat divergence must be rated as low. The uncertainties in this complex area are well summarized by Futuyma and Mayer (1980): "But there are no quantitative data on the genetics of habitat choice, the fidelity of habitat choice, the frequency of encounter of potential mates . . . ." Because different phenotypes may vary in response to a given environment, and conversely a given phenotype may have varying responses in different environments, the study of habitat selection requires studies in both "dimensions" perhaps simultaneously, if the experiment itself does not then become excessively large. Possible designs are considered by Parsons (1967).

Colonization events tend to involve extreme environments, so structural genes of relatively large effect should predominate in the genetic changes resulting from the effects of hard selection on ecological phenotypes (Section 6.3). By contrast, speciation may occur at its maximal level in rather benign habitats, such as humid tropical rain forests where selection is likely to be softer, and where the development of ethospecies in the sense of Carson (1978) may be maximized. On top of major additive genes controlling largely quantitative traits (Templeton, 1980a,b), other types of genes including regulatory genes may be important (Carson, 1977) in the speciation process in benign habitats.

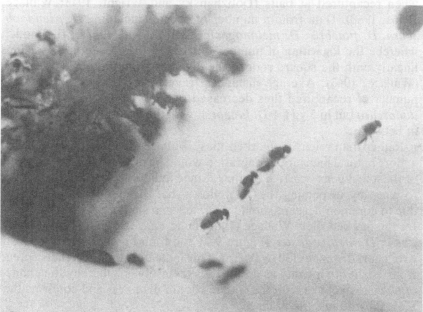

Figure 9.6. (*Top*) Flower of *Hibiscus heterophyllus* in southern Queensland showing *Drosophila hibisci* on the petals. Flies are aggregated in the shaded regions to avoid stress. (*Bottom*) Adult flies with some courtship-type interactions on the petal of *Hibiscus heterophyllus*. (Source: Cook, Parsons, and Bock, 1977.)

It is important to note, however, that experimental evidence for regulatory genes in the speciation process is only just beginning to appear (Dickinson, 1980). In both extreme and benign environments, however, electrophoretic loci should normally be affected indirectly if at all. At the genetic level, therefore, the major difference between the genetic architectures involved in colonization on the one hand and speciation following habitat alteration on the other may be slight. The differences may result from relatively minor adjustments of the genome involved in accurate adaptations to habitats as specialism evolves, whereby each phenotype selects the precise habitat to which it is best adapted. In any case, for both colonization and speciation, additive genes are important determinants of phenotypes at the levels of morphology, physiology, development, behavior, life history, ecology, and so forth.

9.4 Migration and dispersal

Until recently, most studies of migration in *Drosophila* have involved the release of large numbers of flies and then noting their positions when recaptured at baits (Dobzhansky and Wright, 1943; Wallace, 1966a, 1968). Data from a number of species including *D. pseudoobscura*, *D. funebris*, *D. melanogaster*, and *D. willistoni* fit the model, whereby the logarithm of number of marked flies recaptured declines linearly with the square root of the distance from the point of release (Wallace, 1968). Average distances traveled are not great, since the number of recaptured flies decreased by 90% in 100 yd for *D. pseudoobscura* but in 5 yd for *D. funebris*, with the other two species falling in between.

Rather than releasing marked flies, Wallace (1966b) considered that a more natural measure of migration would be obtainable from an assay of the lethal genes carried by flies. Since lethals are carried from place to place by dispersing flies, it follows that the frequency of allelism due to common descent should decrease in a similar way to that found for the marked flies. Wallace (1966b) and Paik and Sung (1969) tested this in *D. melanogaster* (Figure 9.7) and found a decrease in allelism over short distances (30 to 180 m). This indicates that the lethals spread from locality to locality as do the flies themselves, whereby the chance of allelism declines with the square root of distance, and confirms that most flies do not travel far.

More recent experiments using flies marked with fluorescent dust, however, have given rather higher rates of dispersal than these early studies. For example, Powell et al. (1976) obtained a mean dispersal distance of about 170 m/day in populations of *D. pseudoobscura* in the forests of California, and Crumpacker and Williams (1973) obtained similar results in Colorado.

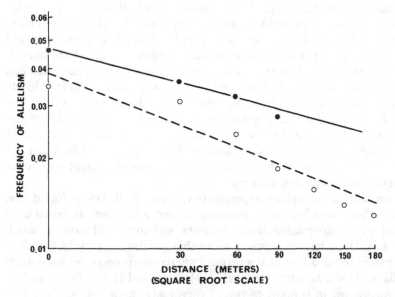

Figure 9.7. The probability (in *Drosophila melanogaster*) that two lethal genomes are allelic as a function of distance between trapping sites at which
they were obtained. [Source: Data from Wallace (1966b) (solid circles) and
Paik and Sung (1969) (open circles) summarized in Wallace (1981).]

The distribution of dispersal distances is, therefore, usually leptokurtic with a maximum close to the point of origin and few individuals
traveling far. On top of this, however, are migrations to new continents,
such as migrations of human groups among continents. In *Drosophila*,
the migration of flies to the Hawaiian Islands and the subsequent migrations between islands form good examples. In this latter case, few
founder individuals may be involved, but the distances are far too great
to be compatible with the leptokurtic dispersal models. Other examples
in *Drosophila* include *D. pseudoobscura* to Colombia in South America
(Section 4.1) and New Zealand (Parsons, 1981a), *D. subobscura* to
South America (Brncic and Budnik, 1980), *D. enigma* from Australia
to New Zealand (Parsons, 1980b), and *D. buzzatii*, which is associated
with the cactus genus *Opuntia* and is now widespread. Inspection of
Figure 3.3 shows that those species that have been tested for tolerance
to physical stresses fall into the cosmopolitan and colonizing range.
Unpublished data of S. M. Stanley and P. A. Parsons on *D. pseudoobscura* are consistent with these conclusions. With time, therefore,
increasing colonization voyages of cosmopolitan and widespread *Drosophila* species might be expected, but a prerequisite is a reasonable
tolerance to physical stresses and the utilization of ethanol as a resource.

In many parts of its range, *D. pseudoobscura* is distributed in a series of large "islands" of suitable habitat separated by deserts or grasslands. In southern California, for example, flies are largely restricted to forested or scrub-covered mountain ranges separated from each other by extensive deserts, but at certain times of the year, flies can be collected in desert oases, orange groves, and other suitable habitats that are scattered throughout the low deserts (Jones et al., 1981). If it could be shown that such populations are periodically extinguished and recolonized from a distant source each year, it would be a result of considerable evolutionary significance. This is because if flies can migrate over long distances to the oases, such migrations might also occur between permanent populations.

From release–recapture experiments, Jones et al. (1981) found that a rather large number of *D. pseudoobscura* will travel as much as 2 km/day over inhospitable desert habitats, and some will travel as much as 10 km. Once having reached a suitable habitat, it is likely that flies will remain there. Jones et al. cite published reports on *D. melanogaster* (Wallace, 1970), *D. nigrospiracula* (Johnston and Heed, 1976), and *D. pseudoobscura* in support of this. Conversely, many of the flies estimated to have traveled 10 km in 1 day could well have migrated further if they had not encountered a suitable habitat. This suggests that migration between mountain and oasis populations is feasible, and between the main montane centers of abundance. Similarly, *D. nigrospiracula*, which is closely associated with its host, the Saguaro cactus, *Cereus giganteus*, frequently migrates between cacti as much as 1 km apart, and migrations up to 15 km appear likely in cactophilic species (Heed and Heed, 1972; Johnston and Heed, 1976). In the tephritid flies (Table 9.4), the colonizing species, *Dacus tryoni*, has been shown to disperse at least 24 km from the center of emergence (Fletcher, 1974) compared with 0.5 km for the habitat-specific *Rhagoletis cerasi* (Bateman, 1977). Comparative studies of colonizing *Drosophila* species with closely related noncolonists would be of obvious interest.

Extensive movement across unfavorable habitats may be important in the maintenance of high levels of electrophoretic polymorphism and in the general lack of association of electrophoretic variants with habitat across the range of *D. pseudoobscura*. Even so, there is some evidence for habitat selection in field populations of *D. persimilis* and *D. melanogaster* (Section 8.2). However, such divergence could be relatively transient perhaps varying with the time of year. In other words, in widespread colonizing species migration could be a real force in promoting generalism and would certainly counter the evolution of habitat selection and specialism. This model of occasional long-distance migration could apply to other populations occurring in islands of suitable vegetation (e.g., *D. inornata* in southern Australia). Species of such ephemeral isolated habitats are likely to be *r* species, where an intimate association with habitat is unlikely. In any case migratory species tend to have higher *r* values than closely related nonmigratory species (Section 2.1).

Migration, therefore, permits the movement of genes from one locality to another, and so the migrated genotype interacts with the new environment. If differential migration of genotypes occurs such that certain genotypes tend to migrate more than others, this is migrational selection (Fisher, 1930; Edwards, 1962). This may occur as a result of environmental heterogeneity (see also Parsons, 1963). In this section, the concern is mainly long-distance migration over potentially stressful habitats. Following from the arguments in Section 5.4, migrational selection could involve the favoring of heterozygosity provided that the migrant population is not excessively small.

9.5 Summary

In some species, there are associations of morphological, physiological, and behavioral traits making up ecobehavioral phenotypes that can be related to habitats occupied. For traits important in such habitat selection, it is expected that stabilizing selection will keep them within fairly narrow limits. However, colonizing species are not likely to be so intimately adapted to their habitats, and directional selection may sporadically occur in response to changing environments.

The species of the Australian *Drosophila* fauna show that as species become less generalist, breeding site specificity increases. In addition, the less specialized and potentially colonizing species may be present in a given habitat for a longer period of time annually than more specialized species, perhaps because of breeding site availability restrictions for the latter group. This is apparently associated with a lesser tolerance of environmental extremes in more specialized species. Species diversities are correlated with the floristic diversity of forests. The limiting extremes are the depauperate heat-stressed faunas of the lowland tropics of northern Australia and the cold-stressed faunas of the uplands in the south. Most of the species occur between these extremes, especially in the floristically complex mesophyll vine forests of northern Queensland.

Although animal speciation involves the development of reproductive isolating mechanisms, ecobehavioral factors involved in adaptation to new habitats are also important, so that an overall ecobehavioral approach is necessary. Given predictable resources, for example, trees or shrubs, specialist host races are possible; in certain perhaps rare sets of circumstances, there is the possibility that host races may evolve to become species. Speciation is less likely where an array of resources are used as in generalist species. Ethospecies may be most likely under the soft selection of relatively benign environments. The development of a genetics of ecobehavioral traits important in habitat selection appears very important for progress in this complex area.

Migration may be of greater significance than most field experiments suggest. Indeed, release–recapture experiments show that *D. pseudoobscura* may migrate considerable distances over inhospitable habitats, perhaps up to 10 km/day. In widespread colonizing species, migration could be a real force in promoting generalism and would certainly counter the evolution of specialism, in spite of evidence for habitat selection in such species.

10

The ecobehavioral phenotype: biological control and domestication

Under domestication, we see much variability, caused, or at least excited, by changed conditions of life. [Darwin, 1859, Chap. XV]

It is remarkable that, in spite of the nonisotropy of the European area of diffusion with its many rivers, seas, and mountains, the process of spread of farming appears to have been relatively smooth and homogeneous. [Cavalli-Sforza and Feldman, 1981:42]

10.1 Chemical and biological control – hard and soft selection

Modern society has come to rely heavily on chemicals to control animal and plant pests, which are colonists that threaten people and their resources. The discovery of antibiotics led to cures of life-threatening infections and also eliminated the need for complex manual sterilizations of operating theaters and clinics. Although hospital costs were accordingly reduced, the microorganisms immediately adapted by the development of resistant strains from their survivors. Cavalli and Maccacaro (1952) described the consequence of such selection in the bacterium *Escherichia coli* using a regime where asexually reproducing colonies were plated on near-lethal levels of chloramphenical, so that selection can be regarded as being relatively hard. Those colonies surviving chloramphenicol did so because of a rare mutational event. In the absence of recombination, each new mutation was, therefore, tested on a particular genetic background that remained intact through subsequent (asexual) generations. Consequently in each selected line, the successful mutations may have conferred high resistance largely through their interactions with the rest of the genome. As long as asexual reproduction continued, these favorable epistatic interactions should persist. Accordingly, in sexual crosses, recombination disrupted these gene combinations that generated resistance, thus revealing that different lines had achieved resistance to chloramphenicol by different genetic routes.

King and Sømme (1958) selected for dichlorodiphenyltrichloroethane (DDT) resistance in *Drosophila melanogaster* in two population cages

175

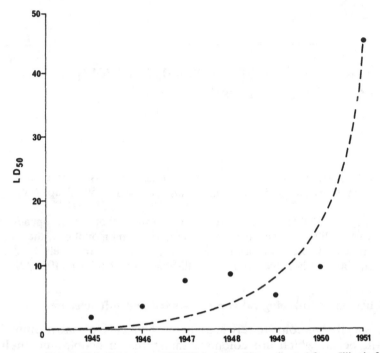

Figure 10.1. Resistance of DDT in houseflies collected from Illinois farms measured in terms of the lethal dose necessary to kill 50% of flies (LD_{50}). (Source: Decker and Bruce, 1952.)

increasing resistance by about 20 times the control over about 50 generations of selection. Genetic analysis revealed that resistance was determined by genes on the three major chromosomes. Crosses between the two resistant strains after 20 generations of selection gave an F_1 population with similar resistance to the parental strain but an F_2 with significantly lower resistance and greater variance. It, therefore, appears that the two strains have achieved resistance by consolidating different combinations of genes for resistance, which argues as with *E. coli* earlier, for resistance being under complex genetic control. Merrell (1965) presented results on populations of *D. melanogaster* in which exposure to DDT had been gradually increased over 10 years by 100 to 300 times the unexposed control, so the changes induced by this form of selection can be very substantial. These results have their parallels in nature. For example, rapid genetic changes have occurred in many insect populations exposed to pesticides, including the effect of DDT on houseflies (Figure 10.1). The same situation occurred in trying to control rats. In England, a gene for resistance to the rodenticide, warfarin, took only 10 years to become widespread in populations of the brown rat, *Rattus norvegicus* (Greaves and Rennison, 1973).

It is likely that a vast number of chemicals affects organisms with different gene combinations leading to resistance in different populations. Antibiotics and insecticides have been given prominence because of their economic importance. A review of *Drosophila* data (in particular Crow, 1957; Parsons, 1980a) shows that as for the majority of quantitative traits, major regions of genetic activity are readily locatable and so can be potentially recombined leading to rapid responses to selection in natural populations. Hence, there are rapid genetic changes analogous to the colonization of natural habitats, which in this case involves the "occupation" of the artificial habitats developed by people to control the organisms in question!

In theory, the aim of insect control by chemical means is to wipe out whole populations. However, given that a more usual result is to build up resistant strains, increasing attention has been paid to biological control measures. There is a considerable literature on the screwworm fly, *Cochliomyia hominivorax*, which is a destructive obligate parasite of livestock, pets, zoo animals, and occasionally people in tropical/subtropical North America (Richardson, 1978). The female mates once, seeks an open wound on warm-blooded animals, lays eggs that become larvae in 12 to 24 hr, followed by soil pupation after 5 to 7 days, and adults emerge between 1 week to 2 months later followed by matings after 5 to 7 days. The times of the life cycle stages independent of the host depend upon climatic factors, in particular temperature and humidity. For adult survival, large temperature fluctuations are detrimental and cool, moist weather is beneficial. There are, therefore, considerable variations of both population size and rate of annual migrations to the north during the warmer months. Since at certain times of the year population sizes are small due to extremes of weather, one control mechanism is to restrict to cold months or hot dry periods human-induced wounds, which would provide oviposition sites for the flies; this is an extremely soft form of selection by comparison with chemical control.

Accordingly, various more effective control mechanisms were sought, the principal one being the sterile male technique. Male flies are sterilized by radiation and released into the environment in sufficient numbers to overwhelm the native population on the premise that sufficient matings between fertile females and sterile males occur to reduce the number of native screwworm flies in the population. Encouragement for this technique came from the elimination of flies from somewhat isolated land masses – Cucaçao in 1954, Florida, and southeastern United States in 1960, and various other Caribbean islands (Novy, 1978). In the major areas of infestation – Texas, Arizona, New

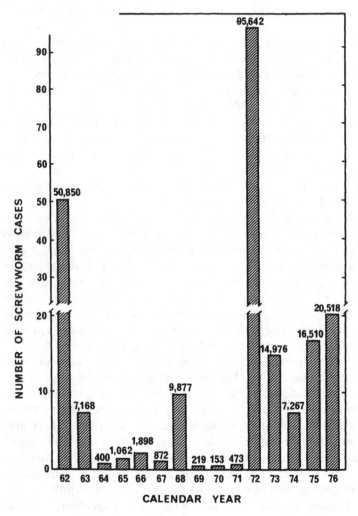

Figure 10.2. Total reported cases of screwworm infestation reported in the United States, 1962–76. (Source: Adapted from Bush, 1978, and Novy, 1978.)

Mexico, and California – releases of "factory-bred" flies began in 1962. The total cases over 1962–76 are plotted in Figure 10.2; high numbers generally occurred when weather conditions were favorable for screwworms, even in the presence of factory-bred sterile flies.

Evidence soon began to accumulate to suggest that the factory-bred flies were becoming less competitive. The whole program is based upon the principle that released sterile male flies must compete with the wild male flies for mates. However, over the years it appears that the factory-bred flies were becoming smaller than wild flies and differed in flight ability and eye color morphism (Bush, 1978), that is, there was a change due to *domestication*. Accordingly, a comparative genetic study of factory-reared and wild populations was carried out.

For the enzyme α-glycerol phosphate dehydrogenase (α-GDH), it was found that almost all factory-bred flies were homozygous α-Gdh_2, whereas the wild Texan flies contained the alternate form α-Gdh_1. Indeed, as soon as new strains were introduced into the factory, α-Gdh_2 began to increase at the expense of α-Gdh_1 (Bush, 1978). The importance of α-GDH is that it is an enzyme involved in flight activity, playing a key role in energy flow in flight muscles, the factory-type enzyme (α-Gdh_2) being less active in the temperature ranges experienced in nature. In the factory, a high-constant temperature was used to speed development, which exerted a strong selective pressure in favor of α-Gdh_2. Hence, in nature, such factory flies could not fly so well as wild flies. In addition, wild flies are active from early morning to late afternoon, whereas factory flies did not arrive on the scene until early afternoon as they were unable to get their flight muscles operating earlier because of a lack of sufficient energy. This means that matings between wild flies are highly likely before the factory flies become active enough for sexual activity. In this case, therefore, genetic changes in the domesticated flies effectively preclude insect control. Domestication has changed a very soft form of selection to almost nonexistent selection.

It demonstrates that a prerequisite for an insect control program is a knowledge of ecology, behavior, and genetics. Indeed, quoting Bushland (1975), who was one of the pioneers in the successful eradication of screwworm in Curaçao and Florida, "there is a great research need for young entomologists to learn much more about screwworm ecology, behavior, physiology and population genetics. Eradication is not as easy as it used to seem." This point was reinforced by Bush (1978). In summary, insect control programs are destined to fail without detailed ecobehavioral information.

The complexity of this problem is well presented by Asman, McDonald, and Prout (1981) when considering genetic control systems for mosquitoes. They summarize numerous studies that they consider to be necessary for the success of genetic control methods in the field situation (Table 10.1). Even so, they are optimistic that although progress may be slow, carefully designed programs may lead to success with mosquitoes. The literature on insect control is large and ever-increasing. Enormous sums are being spent on insect release programs based upon various cytogenetic "tricks," whereby released insects mate with field populations to give inviable or sterile hybrids (Foster et al., 1972; Asman, McDonald, and Prout, 1981). The essential pre-

Table 10.1. *Studies considered relevant to the success of genetic control mechanisms in the field*

A. Release material
1. Successful colonization of species
2. Basic genetic characteristics, depending on method
3. Release material ensured not to be serious as vector
4. Potential genetic mechanism for suppression or replacement
5. Quality mass-rearing
6. Sexing mechanism when only one sex required
7. Method of transport for field release
8. Competitive mating behavior
9. Adequate numbers to achieve desired release ratios
10. Method and time of release
11. Assortative mating, if present
12. Field behavior: dispersal, survival, mating pattern
13. Reproductive physiology in field
B. Indigenous population
1. Population size and fluctuations
2. Density-dependent information, capacities for increase
3. Mating behavior: time, place, female monogamy, male polygyny
4. Life history: resting sites, feeding, dispersal, survival, life cycle, natural enemies
5. Degree of dispersal and immigration
6. Genetic variability within and between populations
7. Geographic differences in mating biology
8. Potential role as disease vector
C. Additional considerations
1. Computer-simulation model
2. Monitoring methods for quality control of release material
3. Monitoring methods for evaluating success of release
4. Cost analysis compared to other methods

Source: From summary in Asman, McDonald, and Prout (1981).

requisites for success encompass the items in Table 10.1. However, as shown for screwworm flies, it is particularly important that the fitness differences between the released and wild insects be as small as possible. Even having effectively demonstrated this last point in an attempt to control a wine-cellar population of *Drosophila melanogaster* based upon the release of individuals homozygous for chromosomal translocations, a small amount of migration was highly disruptive (McKenzie, 1976). Long-term control by genetic means without a detailed initial ecobehavioral assessment is indeed difficult.

A final example concerns biological control of the rabbit, *Oryctolagus cuniculus*, in Australia by the virus disease myxomatosis; both host and virus are colonists in Australia (Fenner, 1965). Myxomatosis is a highly lethal disease of domesticated European rabbits in various parts of South America. The natural reservoir of the myxoma virus is

Table 10.2. *Comparison of the virulence for the rabbit,* Oryctolagus cuniculus, *of field strains of myxoma virus in Australia and Britain a decade after the introduction*[a]

Grade of severity:	I	II	IIIA	IIIB	IV	V
Mean survival times (days):	≤13	14–16	17–22	23–28	29–50	—
Case mortality rate (%):	99.5	99	>90	<90	60	≤30
Australia						
1950–1	100	—	—	—	—	—
1958–9	0	25	29	27	14	5
Britain						
1953	100	—	—	—	—	—
1962	4	18	39	25	14	1

[a] Expressed as percents.   *Source:* Data from Fenner (1965).

the tropical forest rabbit, *Sylvilagus brasiliensis*, in which it produces a benign localized fibroma of the skin. Infection of *Oryctolagus* is due to mechanical transfer of the virus from the skin tumors in *S. brasiliensis* by mosquitoes, and then other domestic rabbits may be infected mainly with arthropod vectors. The European rabbit was introduced into Australia in 1859 and rapidly spread over the southern part of the continent, where it became a pest of major economic significance. The virus was introduced in 1950 to help control rabbits and spread rapidly over the continent with enormous mortality. Insect vectors, principally mosquitoes, were responsible for its spread.

On initial release, the virus caused mortality rates exceeding 99%. The virus spread rapidly from one animal to another during the summer when the mosquito population was abundant. It might be predicted that the disease would disappear after each summer, due to the absence of susceptible rabbits and greatly lowered opportunity for transmission during the winter because of the scarcity of vectors. Although this presumably occurred in some areas, the capacity to survive the winter conferred a great selective advantage on viral mutants able to cause a less lethal disease, whereby a rabbit survived an infection for weeks rather than days. Such mutants appeared within the first year of virus introduction and became dominant within 3 to 4 years (see Table 10.2 for comparison of virulence of field strains). Genetically, the original highly lethal introduced virus has now been replaced by a heterogeneous collection of strains, all of lower virulence.

Rabbits that recover from myxomatosis are subsequently immune, which suggests that selection for genetically more resistant animals should rapidly occur. Indeed, it was found that the genetic resistance of rabbits steadily increased in areas of continued annual exposure

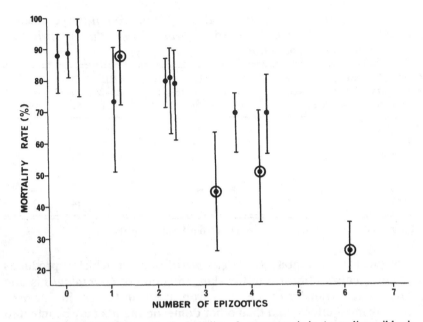

Figure 10.3. Changes in the lethality of myxomatosis in Australian wild rabbits inoculated intradermally with a standard dose of myxomatosis virus of standard virulence. Epizootics occur annually. The data indicate mean case mortality rates and 95% confidence limits. [Source: Data from Marshall and Douglas (1961) presented by Fenner (1965).]

(Figure 10.3). In other words, colonization by the myxoma virus has altered both virus and host, converting what was quite hard selection to a softer form. However, as in the screwworm fly, ecological considerations are relevant, since the disease is less severe under high temperatures, and this would reduce the selection intensity more.

Turning to *Homo sapiens*, the explosive period of European colonization of all other continents had profound effects on the disease patterns in both migrants and indigenes. Europeans took diseases to new populations. There were explosive outbreaks of measles and smallpox in the indigenous peoples of America and Oceania. Australia was settled in 1788 by inhabitants of the northern hemisphere. At that time, the Aborigines numbered about 300,000, and soon after European settlement their numbers began to fall. There are many accounts of shootings, but there are also accounts that smallpox was particularly lethal. Curr (1886) writes: "The principal diseases which have been introduced by the Whites are smallpox, which carried off probably one-third or one-half of the race; and venereal, which, if slower, shows no abatement."

Furthermore, smallpox may well have begun the collapse of the Aboriginal tribal structure: "This disease undoubtedly, besides reducing numbers, affected a great alteration in the condition of the Aborigines generally, and led probably to the breaking up of some tribes – the remnants coalescing for protection from inimical tribes and for the conservation of their common interests" (Brough Smyth, 1878).

Even today, diseases, such as influenza and measles, run a particularly severe course with high death rates among the Aborigines. Although it could be argued that the environment under which the Aborigines live differs from that of the non-Aboriginal inhabitants, the evidence seems clear that diseases that were previously not present tend to run a *much* more severe course than in populations in which such diseases were common for a longer period of time. Additional examples are known for other populations suddenly exposed to diseases new to them (Fenner, 1977). Even Europe may not have escaped, for immediately after the return of Columbus from the New World, some have argued that syphilis ran a very severe course for 50 to 100 years, whereupon it tended to become less severe in effect, presumably because the European population had built up genetic resistance to the disease (Morton, 1966). In other words, diseases caused by microorganisms may devastate a population initially, but eventually there is adaptation as genetically resistant individuals survive and reproduce at the expense of susceptible individuals. In some cases, however, for small isolated populations complete elimination may occur, and one suspects that the recent demise of the Aboriginal Tasmanians may have been partly the result of susceptibility to diseases to which they were not previously exposed.

10.2 Domestication – soft selection

Domestication of plants and animals occurred when the hunter–gatherer food economy was replaced with that of food production by agriculture and animal breeding. Evidence of domesticated animals and plants starts at a date somewhere near the end of the European Ice Age, that is, more recently than 10,000 BC (Ucko and Dimbleby, 1969). The complex set of innovations making up agriculture occurred as a wave of people migrated, namely, the farmers from the Middle East. Using archeological evidence incorporating radiocarbon dates, agriculture took over 3000 years to spread across Europe (Ammerman and Cavalli-Sforza, 1971). It is quite remarkable that in spite of the many rivers, seas, and mountains, the process of spread of farming appears to have been relatively smooth and homogeneous, as shown by the isochrons – lines representing equivalent times – in Figure 10.4.

Another example of homogeneity is the range expansion of the marine toad, *Bufo marinus* (Figure 6.3). It is fascinating that range ex-

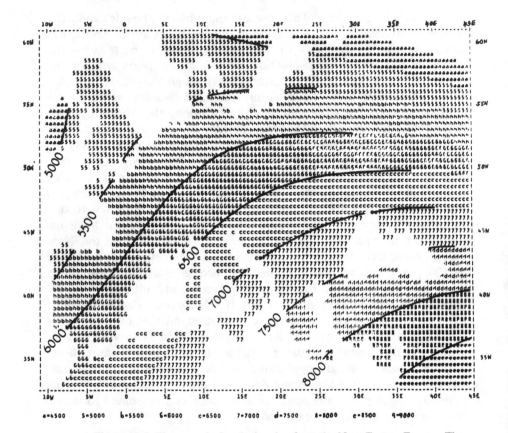

Figure 10.4. The spread of early farming from the Near East to Europe. The lines are isochrons (lines of equal time of arrival) for years before the present and are limited to the range for which archeological (radiocarbon) data are available. [Source: Data from Ammerman and Cavalli-Sforza (1971) and plotted by Cavalli-Sforza and Feldman (1981) with additional data.]

pansions in ecology, archeology, and epidemiology all tend to follow this homogeneous model (Cavalli-Sforza and Feldman, 1981). These similarities follow from theoretical models concerning the wave of advance in situations where processes of diffusion and growth occur simultaneously. Such a model was first put forward by Fisher (1937), who was interested in the problem of the spread of an advantageous gene, and was developed in terms of population growth by Skellam (1951). Fisher's model requires that the square root of area be proportional to time rather than the log of the area plotted in Figure 6.3. I am indebted to L.L. Cavalli-Sforza (personal communication) for pointing out to me that the data of Sabath, Boughton, and Easteal (1981) fit this model except for an outlier (Figure 10.5), which largely dis-

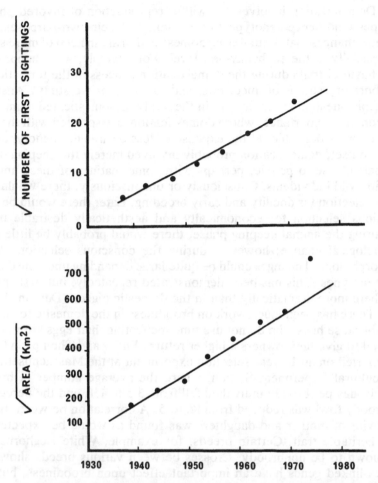

Figure 10.5. *Bufo marinus* data (see also Figure 6.3) plotted according to the theoretical model of Skellam (1951) on the wave of advance(range expansion) of organisms. It should be noted that the cumulative percent of first sightings fits better than the geographical area.

appears when the square root of the cumulative percent of first sightings is used. This measure should be proportional to area and is perhaps more accurate, since *B. marinus* is a large and conspicuous animal found most commonly around human habitation, and when present, easily observed. In any case, public interest in the species has been high since first introduction, and its occurrence in new areas is often discussed and has frequently been reported in newspapers. The appearance of individuals is apparently an event likely to be remembered by local inhabitants (Sabath, Boughton, and Easteal, 1981).

Domestication involves the willful reproduction of favored phenotypes (and hence genotypes) by humans for their own purposes. Genetic changes that occur during domestication are in need of more study especially at the ecobehavioral level. For example, what happens to behavioral traits during the domestication process in the formation of laboratory strains of mice, rats, and rabbits that we study? Answers to this question are important in the study of domesticated animals of economic importance, where domestication is associated with the production of desirable, at the expense of less desirable, phenotypes. At the outset, domestication probably involved merely the keeping of animals of use to people, perhaps with some matings of these animals with wild individuals. Consciously or unconsciously, there would also be selection for docility and early breeding. Later there would be conscious selection for economically and aesthetically desirable traits. During the animal-keeping phase, there would probably be little morphological change; however, during the conscious selection phase, morphological changes could be quite large depending upon the desired phenotypes. This has been demonstrated repeatedly, but perhaps nowhere more dramatically than in the domestic pigeon (Darwin, 1859).

There has been much work on broodiness in the domestic fowl. This is because hens who do not use time incubating their eggs after laying, tend to give their owners a higher return. Fuller and Thompson (1960) reported on an 18-year selection experiment at the Massachusetts Agricultural Experiment Station, where the average number of broody episodes per broody individual fell from 3.5 to 1.1, and the percent of broody fowl was reduced from 86 to 5. A correlation between the behavior of mother and daughters was found as would be expected for a heritable trait. Certain breeds, for example, White Leghorns, are known to be nonbroody. Crosses between various breeds show that sex-linked genes have an important effect upon broodiness, but this may not apply to all breeds.

There are numerous breeds of fowl classified according to place of origin, such as Asiatic, Mediterranean, English, and American breeds (Guhl, 1962). The origin of the domestic fowl is lost in antiquity although four "species" have been recognized in southeastern Asia and India. Crosses between wild jungle cocks and domestic hens regularly occur at least in India (Fisher, 1930). The familiar *Gallus domesticus* could well have been produced by hybridization from several wild species, later becoming distinct from them. This would no doubt have been assisted by fanciers in the early part of this century, who bred fowl for exhibitions based upon colors, plumage, comb characteristics, and range of body weights (see Guhl, 1962).

Among domestic animals, fowl have the most specialized of husbandry procedures; they are maintained indoors in compact groups

Table 10.3 *Behavior fostering adaptation to domestication*

Favorable characteristics	Unfavorable characteristics
1. Group structure	
a. Large social group (flock, herd, pack), true leadership	a. Family groupings
b. Hierarchical group structure	b. Territorial structure
c. Males affiliated with female group	c. Males form separate groups
2. Sexual behavior	
a. Promiscuous matings	a. Pair-bond matings
b. Males dominant over females	b. Male must establish dominance over or appease female
c. Sexual signals provided by movements or posture	c. Sexual signals provided by color markings or morphological structures
3. Parent–young interaction	
a. Critical period in development of species bond (e.g., imprinting)	a. Species bond established on basis of species characteristics
b. Female accepts other young soon after parturition or hatching	b. Young accepted on basis of species-specific phenotype (color patterns, pheromones)
c. Precocial development	c. Altricial development (requiring relatively prolonged parental care)
4. Responses to human beings	
a. Short flight distance	a. Extreme wariness and long flight distance
b. Least disturbed by human ubiquity and extraneous activities	b. Easily disturbed by people or sudden changes in environment
5. Other behavioral characteristics	
a. Catholic dietary habits (including scavengers)	a. Specialized dietary habits
b. Adapt to a wide range of environmental conditions	b. Fixed or unique habitat needs
c. Limited agility	c. Extreme agility

Source: From Hale (1962).

and/or isolated in laying cages. In addition, with trends toward mass production and efficient management, the social behavior of these birds has commanded considerable attention (see Table 10.3 where a list of behavioral characteristics adapting species to domestication is given).

It is to be expected that detailed studies have been carried out on behaviors such as those preceding egg laying (Wood-Gush, 1972). Two strains, White Leghorn and a Brown strain of Rhode Island Red and Light Sussex origin, were studied in battery cages. The Brown strain sat significantly longer than did the White strain, which showed an excessive amount of pacing during the prelaying period. These differences were not affected by enclosing the battery cages and darkening the room. Both strains responded to frustration in a feeding situation by excessive pacing, from which it is apparent that White females were generally upset during the prelaying period. However, some White strain females did sit suggesting the likelihood that selection for sitting in the prelaying period may be successful.

Also, it is not surprising that levels of aggressiveness have been studied in different breeds, since the domestication of fowl is historically related to cockfighting, which suggests early selection for aggressiveness and related traits. For example, fighting cocks have been observed to be shiftier, faster, and less clumsy than domestic cocks, and among varieties of fighting cocks, there are differences in their methods of attack. It is also reasonable then that selection for high and low levels of aggressiveness has been successful in White Leghorns (Guhl, Craig, and Mueller, 1960). A consequence of variable levels of aggressiveness in a flock is the peck order, which is the ranking of individuals according to the number of individuals each flockmate dominates by unidirectional pecking or threatening. The intensity of these contests is greater among cocks than hens. To what extent there are genetic differences among birds in a peck order is at present relatively unexplored, but must be likely. It has been found, however, that males with pea and walnut combs assume lower social positions when intermingled with single comb flockmates (see Siegel, 1979). Isolating birds in laying cages eliminates peck orders, although possibly dominance– subordination relations could occur between birds in adjoining cages. The adaptation of birds to these various procedures is part of domestication, a continuing process that varies in goals in different places and times.

The literature on the behavior genetics of domestic fowl is now quite substantial. Siegel (1979) has reviewed the field from several viewpoints: the process of domestication, behavioral effects on selection, and in the context of genotype–environment interactions [for this latter see also McBride (1958)]. He lists, in addition, a number of mutants with behavioral effects. He stresses the point that poultry breeders have consciously and subconsciously changed behaviors through selection and an interfacing of all of the fowl studies is necessary for the understanding and facilitation of the process of adaptation to the environments in which they are placed.

The Aylesbury duck is a domesticated bird descended from the mallard, *Anas platyrhynchos*; it, therefore, offers the opportunity to make direct comparisons between domesticated and wild forms. Desforges and Wood-Gush (1975a) compared the habituation rates of domestic ducks and wild mallards when a stuffed hen was placed in their home pens and when given novel food in a normal or novel container. The domesticated ducks habituated faster than the mallards. Furthermore, in tests involving escape from a human handler, mallards ran away significantly faster than domestic ducks. It has been suggested many times [see Hale (1962) and Table 10.3] that fitness in a captive environment is enhanced by low reactivity, and because of this the propensity to approach and investigate novel objects would be greater for domestic than wild animals, simply because there would be a high probability of encountering unfamiliar and artificial situations in the environments of domesticated animals. This would seem to be a criterion of importance for selecting animals better suited to a captive environment.

Desforges and Wood-Gush (1975b) compared individual distances between members of flocks of domestic and mallard ducks. Even though the Aylesbury duck is three to five times heavier than the mallard, individual distances (cm) are smaller in the former than the latter whether considering like or unlike sexed pairs while feeding or when resting:

| | Feeding pairs | | | |
	♂♂	♀♀	♂♀	Resting
Domestic	30.5	—	—	77.5
Mallard	45.7	30.5	17.8	129.5

This indicates that during domestication there has been selection against aggression and for reduced individual distance. Since the domestic fowl has been domesticated for longer than the Aylesbury duck and has been selected for adaptation to an intensively agricultural environment, it would be of great interest to compare domestic and wild jungle fowl in this way.

A final set of comparisons concern sexual behavior (Desforges and Wood-Gush, 1976). There have been changes in intensity of the social displays of the Aylesbury duck, which tend to reduce the attention-catching features of these displays, and the down–up display is performed less frequently than it is by mallards. During domestication, there is no premium upon promoting social displays that partially function as interspecific ethological isolating mechanisms, since in the environment of domestication the possibility of hybridization disappears, so that biologically the situation is similar to geographical isolation in the wild. In addition, female Aylesbury ducks incite several drakes, whereas mallards incite only one, so that mallard ducks only form pairs. Although copulatory behavior tends to be promiscuous in Aylesbury ducks, it is not completely so, because males and females direct most of their sexual activity to only two or three individuals. As Hale (1962) and others have pointed out, it is of economic importance that domestic animals should at least tend to be promiscuous as in these ducks.

Table 10.3 gives a list of behavioral characteristics adapting species to domestication as summarized by Hale (1962). The genetic study of many of these would be particularly rewarding, for example, those traits already cited of importance in the domestication of fowl and ducks. Biometrical experiments to assess whether behavioral traits are predominantly subject to directional selection in nature or to stabilizing selection will be important in attempting to assess the likelihood of a domestication goal being attained. There are some parallels with the generalist ecobehavioral phenotypes involved in colonization, since domestication and colonization involve adaptation to novel habitats – in the former case deliberately artificial, and in the latter often unintentionally artificial. For example, the catholic dietary habits and adaptation to a wide range of environmental conditions (Table 10.3) involved in domestication has direct parallels in colonization. Since the changes discussed in this section are largely behavioral, soft selection is largely involved; it would be unusual for whole populations to disappear during a domestication process.

10.3 Summary

Antibiotics lead to cures of life-threatening infections; however, antibiotic-resistant strains of microorganisms have been rapidly built up. The same occurs for insecticides, where resistance is controlled by potentially locatable genes that can be recombined leading to rapid responses to selection in natural populations. In recent years, increasing attention has been paid to biological control measures. The sterile male technique has been successful in a limited way, but in many cases there is a lack of solid ecobehavioral information on the target organism, which is restrictive. Myxomatosis is an excellent illustration of biological control, where the severity of the disease fell after a few years since both virus and host (the rabbit) were modified genetically converting what was relatively hard selection to a softer form. Colonizing microorganisms, whether in animals or humans, may therefore offer opportunities for the study of evolution in action.

Domestication involves the willful reproduction of favored phenotypes by humans for their own purposes. This phase of history began somewhere near the end of the European Ice Age when the hunter–gatherer food economy was replaced by food production by agriculture and animal breeding. In organisms such as fowl and ducks, there have been extensive morphological and behavioral changes to obtain desired phenotypes. There are some parallels with the generalist ecobehavioral phenotype involved in colonization. This is because domestication and colonization involve adaptations to novel habitats – in the former case deliberately artificial and in the latter case usually unintentionally artificial.

11

Parasites and plants

Lewontin: Yes, in other words, you are always colonizing, but you grow so large so fast that you knock yourself out. Mayr: Your arc goes sky high. Lewontin: There is a sort of parasitic ability to colonize. [Discussion at end of Lewontin (1965:94).]

Successful colonizing species are distributed in many different families and they represent great diversity with respect to morphological and physiological characteristics. There is, however, one feature which the great majority of these notably successful colonizers share: a mating system involving predominant self-fertilization. [Allard (1965:49) writing on plants.]

11.1 Parasites

About three-quarters of the known animals on earth are insects. Of these, at least 60 % are parasites (Table 1.1). An estimate that parasitic insects represent close to half the animals on earth does not, therefore, seem unrealistic (Price, 1980). Adding to these the large groups of parasitic animals found among the nematodes, flatworms, mites, and protozoa, it is clear that the parasitic way of life is more common than all other feeding methods.

Parasites have a negative impact upon their hosts, which act as the environment for the parasite, as repeatedly pointed out by parasitologists and others. The host reacts defensively and individually over the short term to the impact of the parasite (e.g., immune responses) and over the long term as a population coevolving with its parasite population. Therefore, at the genetic level coevolutionary phenomena are very important. Short generation times and large populations of parasites compared with the hosts permit a very precise tracking by the parasites of the genotypes controlling resistance in the host. After the evolution of new host defenses, escalation of attack by parasites is likely to proceed rapidly. Exceedingly dynamic coevolutionary systems are likely to occur as demonstrated in Section 10.1 for the myxoma virus–rabbit interaction.

Parasites must be put into the context of this book even if briefly,

191

because the parasitic way of life means the continual colonization of hosts, which exist in a matrix of inhospitable environments. Since resources are likely to be widely scattered and very patchy in distribution, the colonization of new hosts is hazardous (Price, 1980). Adaptations by which this hazard is reduced include mass production of spores and eggs, dispersal of inseminated females that form a high proportion of the population, dispersal by attachment to a larger species (often the host), and extreme longevity of resting stages. Some of these characteristics also occur in patchily distributed colonizing species of plants.

In many cases then, single females may found a colony on a resource remote from its origin. It is not, therefore, surprising that reproductive systems in parasites are much more diverse than for the colonizing animal species that have been discussed in this book. Such reproductive systems include inbreeding among progeny, hermaphroditism, and asexual reproduction, including polyembryony and parthenogenesis. Among parasites, parthenogenesis is particularly common.

The adaptive features of parthenogenesis are numerous (Williams, 1975; Price, 1977, 1980) for example:

1. A single female can establish a clone so that multiplication can occur when there is little or no chance of contact occurring between more than one individual of the same species.

2. In a patchy, complex, and changing environment, parthenogenetic organisms may be maladapted, because of the uniformity of the progeny of a single female compared with the progeny of females producing fertilized eggs. This disadvantage may, however, be reduced by the utilization of highly stable and predictable microenvironments as occur in *living* host organisms, which provide a high degree of homeostasis. Price (1977) argues that positive feedback may reinforce the evolution of parthenogenesis in parasitic organisms (Figure 11.1).

3. The gross reproductive rate of parthenogenetic females is virtually doubled, so that the probability of finding a new host is increased.

4. Adaptive combinations of genes can be fixed without danger of disruption (White, 1973). This could be of importance in the close coevolutionary tracking of the host system.

Although some authors regard parthenogenesis as a dead end in an evolutionary sense, Price (1980) after extensively reviewing the situation points out that this may not be so. Considerable variability has been found in parthenogenetic species, so providing ample material upon which natural selection can act. In addition, the short generation time and high fecundity of parasites mean that a large number of in-

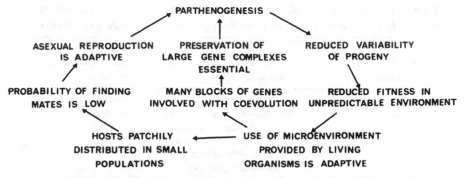

Figure 11.1. Positive feedback loop model for the evolution of parthenogenesis among parasitic organisms. (Source: After Price, 1980.)

dividuals are produced per unit time relative to hosts or free-living organisms of a similar size. Thus, mutations could provide enough variability to permit long-term evolution of a parthenogenetic species.

In summary, the general pattern envisaged for parasitic species includes relatively homozygous populations, with little gene flow between populations (Price, 1980). This means many specialized races, rapid evolution, much speciation, and many sibling species. Important contributory factors to the evolutionary potential of parasitic species include short generation times, large fluctuations in population size, and narrow tolerances to environmental conditions. Price (1980) provides a tabular summary (Table 11.1) of the kind of species a parasite is likely to be by comparison with the free-living, relatively large organisms that have formed the main thrust of this book.

It is important to note (Grant, 1975) that in other small organisms having short generation times, asexual reproduction tends to play an important role, for example, asexual fission in bacteria and the protozoan *Paramecium*, parthenogenesis in the planktonic crustacean *Daphnia*, and the insect *Aphis* (Section 8.4), and self-fertilization in *Paramecium* and the land snail, *Rumina decollata* (Selander and Kaufman, 1973). Interestingly, the inbreeding forms of paramecia have a shorter generation time than the outbreeding forms (Sonneborn, 1957) which tends to occur in higher plants. All of these organisms have colonization episodes mainly involving the asexual phases, so that as in parasites, populations are capable of rapid expansion to exploit ecological opportunities. In fact in insects, the evolution of parthenogenetic races and species has often been associated with the colonization of areas unoccupied by ancestral bisexual forms (White, 1973). For example, in the cockroach, *Pycnoscelus surinamensis*, 10 clones were identified on the basis of variation at five enzyme loci deriving from the bisexual ancestor *P. indicus* (Parker et al., 1977). These clones are

Table 11.1. *A tentative description of the kind of species parasites are likely to be compared to free-living, relatively large organisms*

	Kind of species	
Criterion	Parasites	Predators
1. System of reproduction	Biparental sexual reproduction to mitotic parthenogenesis with a strong tendency toward hermaphroditism and parthenogenesis	Biparental sexual reproduction
2. Amount of gene flow	Largely inbreeding	Outbreeding
3. Size of populations (demes)	Usually small to very small but erupting to very high numbers at times	Large and fluctuating to a lesser extent
4. Phenotypic plasticity	Polymorphic ranging to monomorphism even between species (siblings)	Monomorphic, without sibling species
5. Sequence of generations	Usually rapid	Slow
6. Environmental tolerance	Narrow	Broad
7. Difference in origin	Rapid by sympatric (gene mutations) or parapatric (chromosomal mutations) speciation; instantaneous by polyploidy	Slow by allopatric speciation
8. Rate of evolution	Rapid	Slow
9. Structure of species	Often polytypic	Monotypic
10. Variation in chromosome number	None in many species to much	Little variation or none
11. Degree of intra- and interspecific fertility	Great range in intra- and interspecific fertility	High intraspecific, low interspecific fertility
12. Presence or absence of hybridization in nature	Interspecific hybrids found rarely	Usually none
13. Pattern of distribution	Very local to widespread	Widespread

Source: Price (1980).

highly successful invaders. Presumably, natural selection favors parthenogenesis in ecological settings where there is a premium upon high fecundity and rapid dispersal into the predictable habitats favored by cockroaches. Genetic diversity among flexible clones may further increase the probability of the colonizing success of a parthenogenetic race over a sexual population in areas where there are several habitat types.

Given the stress on coevolutionary phenomena, it would appear that explorations into the evolutionary biology of parasites could be profitably carried out using groups of host organisms for which there are sophisticated studies of their own evolutionary biology. For example, parasitism of Drosophilidae by wasps has often been recorded (Basden, 1972). For the parasite *Leptopilina heterotoma* (= *Pseudocoila bochei*) (Hymenoptera:Cynipidae) of *Drosophila melanogaster*, there are basic data on life cycle, fecundity, and behavior, although most interest has centered upon the defense reactions of the host (Salt, 1970 for review) rather than upon the biology of the wasp. An important discovery was made by Prince (1976), who found a braconid wasp *Asobara* (= *Phaenocarpa*) *persimilis* (Braconidae: Alysiinae) that parasitizes the larvae of several *Drosophila* species in Victoria, Australia. The Alysiines are all endoparasites of Diptera (Capek, 1965) and *A. persimilis* is the first described member of the subfamily native to Australia.

There appears to be several levels of host finding in *A. persimilis* relying upon different types of stimuli. For example, there is a long-distance effect involving smell, attracting females to sites (e.g., rotting fruit) where *Drosophila* larvae are likely to be found. Hundreds of female wasps, but no males, have been caught in fermented-fruit traps in the field; however, aspirator catches in the same area suggest that the field sex ratio is close to 1:1. Short-distance effects appear to involve several types of stimuli (Prince, 1976). As one example, female wasps are strongly attracted to dead yeast medium, as shown by the searching patterns in Figure 11.2.

When a host larva is found, the brown ovipositor of the wasp is freed from its black sheath, and within 1 to 2 sec of inserting the ovipositor, the larva becomes immobilized. The larva recovers in 60 to 90 sec and shows no obvious ill effects. For *D. melanogaster* and *D. simulans* larvae in one series of experiments, the mean oviposition time was 18.6 and 18.5 sec respectively. In *D. melanogaster*, there were no significant differences in oviposition rates between larval age groups ranging from 8 to 80 hr. Under experimental conditions, parasitism is very efficient since oviposition rates varied between 78% and 83%. However, when larvae of mixed ages, 8 and 80 hr, were supplied, there were many more contacts with the older larvae presumably because these larvae are bigger, move further, and so are more often within the contact range of the female wasp.

Most entomophagous wasps can recognize early parasitized hosts and will not oviposit in those already containing eggs (Salt, 1970). Prince (1976) took around one hundred 1-day-old larvae for various

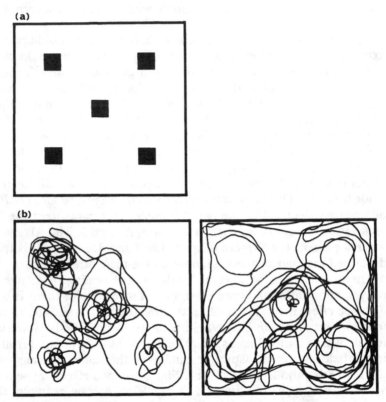

Figure 11.2. Searching patterns in *Drosophila* wasps. (a) The agar medium was divided into squares, and the five marked squares contained dead yeast. (b) Searching patterns of two females of the wasp *Asobara persimilis* on the medium in a. The wasps are clearly attracted to the dead yeast squares. (Source: Adapted from Prince, 1976.)

periods of time, and the number of eggs in each larva was counted. The five cases in Table 11.2 show that there was a highly significant deviation from a distribution of parasitized eggs assuming a Poisson distribution. Thus, the wasp can recognize and will avoid previously parasitized hosts, whether parasitized by the same or a different female.

Considering fecundity, Prince found that 15 wasp females under optimal conditions produced 3852 offspring, giving an average fecundity of 257 per female. He comments that this is comparable with some of the higher values in the literature for such parasites. Further experiments demonstrated that when a wasp egg is laid it is almost always fertile, since development of parasitized and normal *Drosophila* larvae is equally successful.

In summary, the wasp has a mechanism of long-distance attraction

Table 11.2. *Avoidance of superparasitism of* Drosophila *larvae by the wasp* Asobara persimilis

Host species	No. of larvae	No. of eggs	Proportion parasitized	No. of eggs per larvae		
				0	1	2
D. melanogaster	97	34	0.35	63	34	0
D. melanogaster	102	43	0.42	59	43	0
D. melanogaster	88	51	0.58	37	51	0
D. melanogaster	105	86	0.82	19	86	0
D. simulans	101	78	0.77	23	78	0

Source: Data from Prince (1976).

to probable larval sites and a well-developed pattern of nonrandom searching so that a searching female spends much time in areas of high host density. Various ages of larvae are acceptable for oviposition, although when given a choice, older larvae are more often attacked. Large numbers of fertile eggs are produced by females, and they are conserved by strong avoidance of previously parasitized larvae. The handling time spent by the wasp in dealing with each prey is short compared to the potential life span. In addition, development occurs with equal success in a range of host ages. The potential of this wasp to parasitize a high proportion of hosts in a population is clear, and levels of parasitism in the field have been recorded varying from 0% to nearly 80% (Prince, 1976). Diversity in the natural environment may allow a greater proportion of prey to avoid parasitism than indicated by the laboratory results, and as for the hosts, weather, resources, and other environmental factors may affect the behavior and survival of the wasps. Even so, the wasp is undoubtedly very efficient as a "colonist" of its host.

Drosophila melanogaster and *D. simulans* are the most common hosts in Victorian orchards. Comparative parasitization studies of these two sibling species are consistent with the interpretation that in the field, they are parasitized in proportion to their relative abundance. However, *Asobara persimilis* is probably a generalist for species parasitized, since in the laboratory these wasps have additionally been grown in *D. hydei* and *D. busckii*, the Australian endemics *D. fumida* and *D. nitidithorax*, as well as *Scaptomyza australis* (*Scaptomyza* is a genus taxonomically close to *Drosophila*). Curiously, *A. persimilis* does not grow in *D. immigrans*; eggs are readily laid in the larvae but do not develop. Since the possible host list contains representatives of several *Drosophila* subgenera and a species of *Scaptomyza*, *A. persimilis* obviously possesses a fairly general mechanism of inhibition of the host's immune reaction. This ability, plus the relatively non-discriminatory method of host recognition, may help to explain how this endemic parasite has been able to switch so effectively to using recently imported cosmopolitan *Drosophila* species as hosts, in particular *D. melanogaster*. The development of an immune reaction in *D. immigrans* appears inconsistent with parasitization of *D. hydei* of the same subgenus (*Drosophila*). However, *D. immigrans* is a rare inhabitant of rain forests. Despite comments in Section 9.2, this species could perhaps have spread into urban areas from rain forests, so that there could have been time for this species to build up an immune reaction to *A. persimilis*.

Studies on parasites of *Drosophila* have escalated recently, for example, Carton (1977) has studied the olfactory responses of a *Cothanopsis* species (Hymenoptera: Cynipidae), which is a parasite of larvae of *D. melanogaster* collected in the West Indies. Predictably the wasps are attracted to odiferous substances coming from fermenting synthetic medium; ethanol was found to be particularly attractive. Other recent observations on parasites of the same family include those of Chabora, Smolin, and Kopelman (1979). It is of considerable evolutionary interest that Carton and Kitano (1981) have recently found that seven species of the *melanogaster* subgroup can be classified according to cellular immune reaction and success of parasitism of the cynipid wasp *Leptopilina boulardi*. This classification parallels the phylogenetic relationship of these species based upon chromosome banding sequences, although *D. melanogaster* and *D. simulans* are somewhat more different than expected in this criterion, perhaps due to character displacement.

The richest *Drosophila* fauna in Australia is in the humid tropics of northern Queensland in complex mesophyll vine forests (see Section 9.2), and it is a reasonable hypothesis that there should be a parallel fauna of parasitic wasps. A preliminary first survey (P. C. Chabora and P. A. Parsons, unpublished data) revealed 10 species of wasps belonging to three families, Braconidae, Cynipidae, and Diapriidae, 9 of which are new. The tenth was *A. persimilis*, the wasp discovered by Prince (1976) in southern Australia. In addition, he has found this species in the North Island of New Zealand (G. J. Prince, unpublished data). This particular species, therefore, is apparently an extremely widespread generalist species, since it is almost certain that its host(s) in northern Queensland rain forests differs from those in the south. The further study of the hymenopteran parasites of *Drosophila* should be rewarding both in the study of parasite–prey relationships and in considerations of colonization itself. Some of the hypotheses developed and discussed by Price (1977, 1980) would appear to be capable of being rigorously tested.

11.2 Plants

Like parasites, there is an array of reproductive systems in successful plant colonizers. Mayr (1965), in summarizing the 1964 conference on The Genetics of Colonizing Species, comments that events such as polyploidy, hybridization with another species, introgression, reduction in chromosome number, and shifts from perennial to annual habit may be important to varying degrees. One or several such genetic events often seem to be the starting point of a very aggressive phase of colonization. The literature on plant colonization is large and ever-expanding. Here, only a few examples are cited to provide a comparison with animal colonists, although examples mainly involving variability at the intraspecific level have already been cited where they complement animal studies. To do full justice to the plant literature would require another book.

Although Baker (1965) has provided a list of characteristics of the ideal weed, there are considerable variations among such colonizing plant species. Jain (1979) compared six grassland annuals for several variables related to genetic structure and colonizing ability. These are summarized as rank orders in Table 11.3. For example, rose clover, *Trifolium hirtum*, has low rates of seed dispersal but occasional long-distance dispersal with the aid of grazing animals in pastures, as well as utilization by humans. New colonies can be founded that are capable of surviving long periods of time because of high seed output per plant and long-term seed storage in the soil. Wild oats, *Avena*, and the bromes, *Bromus*, differ in their dispersal and reproductive patterns from rose clover, whereby seed dormancy is unimportant, but a plastic response to the environment permits germination over a wide range of conditions. Compared with rose clover, seedling survival is higher in the bromes, and in some circumstances, large stands are rapidly built up. This shows that although there may be an ideal weed, there are numerous and differing colonizing processes among such plants.

A related issue concerns the system of reproduction. As in parasites, asexual reproduction and inbreeding are claimed to be advantageous for their efficiency as well as for genotypic constancy in stable environments. Accordingly, sexuality may be favored when colonization of new environments requires dispersal and strong selection for unique gene combinations (Williams, 1975). The implication here is that the environments are new in an ecological sense, so that evolutionary challenges are likely. Allard (1965) considers that many plants that are conspicuously successful as colonizers are predominantly self-pollinated species. Analyses of the genetic systems of certain of these colonizing species indicate a compromise between the high recombina-

Table 11.3. *Rank orders of estimated means of some variables describing population structure, variation, and numbers in six grassland annuals*

Variable category	Annual grasses[a]				Forbs[a]	
	A	B	C	D	E	F
I. Distribution of species						
a. Geographical range	2	3	1	6	5	4
b. Continuity	2	3	1	5	4	6
c. Local patchiness	5	4	6	1	3	2
d. Patch stability	6	6	6	3–4	1–2	?
e. Presence in weedy habitats	5	2	6	4	3	1
f. Within-patch plant density	6	6	6	1	2	3
II. Gene flow and neighborhood size						
a. Range of seed dispersal	4	3	6	1	5	2
b. Range of pollen dispersal	3	4	5	6	1	2
c. Mean outcrossing rate	5	4	3	6	2	1
d. Effective neighborhood size (N_e)	3	4	2	1	5	6
III. Temporal aspects of species abundance						
a. Life-cycle components						
1. Percent seed carryover	4	3	6	5	2	1
2. Percent survival to maturity	4	5	1	2	3	6
3. Seed output per plant	4	3	6	1	2	5
b. Overall density response						
1. Mortality type	4	3	5–6	5–6	2	1
2. Plastic type	3	4	2	1	5	6
c. Temporal stability in N_e	6	3	3	5(?)	?	?
IV. Genetic variation						
a. Between sites, within regions	4	3	2	6	5	1
b. Between neighborhoods or patches	2	4	5	6	3	1
c. Within patches						
1. Heterozygosity	5	3	4	6	2	1
2. Polymorphism index	5	4	3	6	2	1
V. Phenotypic plasticity						
a. In nature	2	4	1	3	5	6
b. Controlled environments	3	4	1	2	6	5
VI. Variance in fitness and indices of evolutionary potentials						
a. Heritable variation in seed output	5	5	5	5	4	?
b. Reproductive surplus	3	4	5	1	2	6
c. Genetic components of mortality rates	Unknown					

tional potential of outbreeding species and stability expected of self-pollinated species. Indeed, Jain (1976) in a review of evolutionary ideas on inbreeding in plants brought out a number of alternative hypotheses, none of which could be unconditionally supported by existing examples. Price and Jain (1981) surveyed the British flora and found no overwhelming evidence for an association between longevity and colonizing ability, although predominant selfing was found to be more common among colonizers. They caution, however, that such evidence for associations among genetic and ecological factors accounting for high colonizing ability is difficult to interpret when the role of longevity, ploidy, interfamily, and interpopulation genetic variability, uncertainties about habitat classification, and so forth are taken into account. As in animals, therefore, analyses of the complexity of that in Table 11.3 seem necessary since differing evolutionary outcomes occur both intra- and interspecifically.

An observation at the "community" level concerns the sex forms of the South Australian flora divided into introduced (i.e., arrival within 120 years) and indigenous components, compared with a classification by Lewis (1942) of the English flora. There are three basic sex forms in flowering plants, and some species may be combinations of more than one of the three. Most flowering plants are hermaphroditic, having both sexes in the same flower. A species that has separate male and female flowers on the same plant is monoecious, and a species with male and female flowers on different plants is dioecious. Using the flora of South Australia (Black, 1922–52), the species of the South Australian flora can be classified into these sex forms (Table 11.4). A comparison of the three sex forms of the indigenous South Australian flora with the English flora indicates a somewhat greater proportion of dioecy in the South Australian flora (Parsons, 1958), as also occurs in the flora of the southwest of Western Australia (McComb, 1966).

However, there are very few dioecious species in the flora introduced into South Australia within the last 120 years. Indeed the incidence of dioecy is lower than recorded in different floras covering many regions of the world (Bawa, 1980). The introduction of dioecious species requires two separate plants. In addition, Bawa (1980) considers that most dioecious species are insect pollinated. This means that polli-

---

Table 11.3 (*cont.*)

a Species: A, *Avena barbata;* B, *A. fatua;* C, *Bromus mollis*, D, *B. rubens;* E, *Trifolium hirtum;* F, *Medicago polymorpha*. The ranking of means assigns the lowest scales to the highest or greatest value for the variable considered; the highest value is the lowest or least value. A question mark is used whenever the ranking for the species in question is too poorly known for any estimate to be made.
Source: Jain (1979).

Table 11.4. *Distribution of sex forms in the species of the South Australian flora (introduced and indigenous classified separately) and the English flora*

	South Australian flora[a]		
Sex form	Introduced	Indigenous	English flora[a]
Hermaphrodite	445 (92.0)	1871 (89.0)	2080 (92.2)
Monoecious	33 (6.8)	123 (5.8)	122 (5.4)
Dioecious	3 (0.6)	88 (4.2)	54 (2.4)
Mixed	3 (0.6)	20 (1.0)	— (—)
Total	484 (100)	2102 (100)	2526 (100)

[a] No. of species with percentages in parentheses.
Source: Data from Parsons (1958).

nation could be difficult in introduced dioecious species, although many dioecious species of temperate regions are wind pollinated. Therefore, the low incidence of dioecious species in the introduced flora of South Australia is expected. Dioecy, thus, appears to be a property of a fairly static flora and is certainly not a feature of recent colonizations where self-compatibility would be favored especially in weeds, where opportunistic range expansions are the normal expectation (Baker, 1955). The introduced flora ranges across many genera since there are 1.7 species per genus in the introduced flora, 3.7 in the indigenous flora of South Australia, and 4.3 in the English fauna. Thus, plant introductions may be in taxonomically quite diverse groups, especially since the number of genera per family is 4.9, 5.9, and 5.7 for the three categories, respectively. The low species numbers per genus can presumably be accounted for by the absence of noncolonist species in the various introduced genera.

The approach outlined in Section 3.2 of comparing colonizing species with closely related noncolonists has been quite widely advocated in plants (Baker and Stebbins, 1965; Harper, 1977) and permitted Baker (1965) to attempt to define an ideal weed. The array of systems of reproduction in plants, however, makes this task difficult compared with the majority of animals upon which relevant studies have been carried out; animals with phases including asexual reproduction, in particular, parasites, are a class with equivalent difficulties.

Another approach is that outlined in Section 6.1 where genetic variability levels are compared among species, in particular widespread species. Jain (1979) compared patterns of phenotypic plasticity and genetic variation in three such pairs of congeners *Avena fatua* and *A. barbata, Bromus mollis* and *B. rubens,* and *Limnanthes alba* and *L. floccosa.* Within each of these pairs (Table 11.5) the first-named species

Table 11.5. *Comparisons between related species for several measures of phenotypic plasticity and genetic variation*

Species pair	Species with higher values						
	Total phenotypic variation in nature	Between-families variance	Within-families variance	Response to within-family selection	Induced phenotypic variation under different environmental conditions	Plastic response to density	Allozyme variation within populations
Bromus mollis vs. *B. rubens*	*B. rubens*	*B. mollis*	*B. mollis*	*B. mollis*	*B. rubens*	*B. rubens*	*B. mollis*
Avena fatua vs. *A. barbata*	*A. barbata*	*A. fatua*	*A. fatua*	*A. fatua*	*A. barbata*	*A. barbata*	*A. fatua*
Limnanthes alba vs. *L. floccosa*	*L. floccosa*	*L. alba*	Same in both spp.	No data	*L. floccosa*	*L. floccosa*	*L. alba*

Source: Jain (1979).

has more or less ubiquitous genetic polymorphism, whereas the latter has widespread monomorphism. Evidence from quantitative genetic studies shows that although the phenotypic variance for a majority of characters in monomorphic populations is as large or larger than in polymorphic species samples, the genetic component is small as shown by an analysis of responses to selection and the partitioning of variances within and between families. In these species, therefore, phenotypic plasticity appears to replace genetic variation. Using a series of macroenvironments varying in soil type, moisture, and photoperiod regimes, environmentally induced variability, which is a measure of phenotypic plasticity, was shown to be greater in these species. There are, therefore, no levels of genetic polymorphism typical of colonizing plant species. The answer, if obtainable at all, resides in a total analysis of the type presented in Table 11.3 of plant species and the habitats that they occupy, paying particular attention to those ecologically important phenotypes enabling direct adaptation to habitats.

It is of interest to note that Selander and Hudson (1976) consider that the population structure of the land snail, *Rumina decollata*, in southern France is similar to the strongly selfing grasses, in particular the wild oat species, *Avena barbata*. This parallelism is reasonable since Selander and Hudson (1976) "visualize *Rumina decollata* as a colonizing species of very sedentary habit adapting to temporal and spatial patterns of environmental heterogeneity by the maintenance of multiple monomorphic genotypes, each suited to a distinctive, relatively wide range of conditions." In other words, analogous results may occur in plants and animals, when there are parallels in life-history characteristics.

In conclusion, quoting from a Congress symposium paper from the *Proceedings of the Second International Congress of Systematic and Evolutionary Biology* (Brown and Marshall, 1981):

> The available evidence is contradictory. The relative impact of evolutionary forces acting on colonizing species seems to vary with species and events such that each case is unique. Colonizing species are of interest not so much because they are a homogeneous group, but because they display a wide range of possible evolutionary outcomes.

11.3   Summary

Parasites have a negative impact upon their hosts, which act as the environment for the parasite. This is a highly stable and predictable environment providing a high level of homeostasis. The colonization of hosts is hazardous, and there are a number of adaptations by which this hazard is reduced, in particular, a diversity of reproductive systems, including parthenogenesis.

Exceedingly dynamic coevolutionary systems are likely between parasite and host. Short generation times and large population sizes of parasites compared with hosts may permit a very precise tracking by the parasites of genotypes controlling resistance in the host. Explorations into detailed evolutionary systems of parasites should be carried out using host organisms for which there are sophisticated studies of their evolutionary biology. The hymenopteran parasites of *Drosophila* have considerable potential in this regard, since it is highly likely that there is a widespread and complex hymenopteran fauna paralleling the *Drosophila* fauna.

In plants, as in parasitic insects, there is an array of reproductive systems in successful colonizers. This makes the situation considerably more complex than for the animal colonists that form the main thrust of this book, to the level that the evolutionary outcome in each colonizing plant species appears almost idiosyncratic. However, many plants that are conspicuously successful as colonizers are predominantly self-pollinated. In agreement, the introduced flora that has flourished in South Australia in recent times contains very few dioecious species.

12

Discussion and conclusions

There is no such thing as a sharp separation between genetics,
ecology, the study of behavior, and still other ways of looking at
colonizers. I am sure every ecologist here realizes that he really
ought to know more about genetics, that in a way he should consider the genetics of his organisms as part of their ecology. In a
similar fashion, the geneticists realize that they cannot truly understand many of the genetic phenomena accompanying colonization
unless they study them on the background of the ecological situation, again demonstrating there is no clear division between the
two fields. In animals the study of behavior is quite clearly, in all
cases, a third dimension to this same picture. [Mayr, 1965:561-2]

Too often, the adaptationist programme gave us an evolutionary
biology of parts and genes, but not of organisms. [Gould and
Lewontin, 1979]

12.1 The organism as the unit of selection

The overriding assumption of this book is that the organism is the unit
of selection. The organism can be considered as a behavioral phenotype, ecological phenotype, metabolic phenotype, electrophoretic
phenotype, biochemical phenotype, physiological phenotype, morphological phenotype, and so forth according to the emphasis. Since this
book is on colonizing species and their habitats, the ecobehavioral
phenotype including life-history components dominates. The compound term *ecobehavioral* emphasizes the difficulty of separating ecological and behavioral components in the determination of habitats.

The real world is much more complex than envisaged by those who
consider that evolution is merely a question of how the best alleles in
all loci come to be assembled. Indeed, the real world more follows the
following situation (Wright, 1980):"From the theoretical standpoint,
the continually branching chains of biochemical and morphogenetic
steps that must usually intervene between a primary gene product and
an observed phenotypic effect, imply practically universal pleiotropy."

206

Other authors stressing the universality of pleiotropy include Caspari (1952), Gruneberg (1963), and Lande (1980).

Therefore, the ecobehavioral phenotype is the end product of many genetic pathways interacting among themselves. Adding the complication of the environment, genotypes may differ in response to a common environment, and conversely one genotype may vary in response to different environments, as described in Section 7.2 for the effect of development temperature upon different life-cycle stages among iso-female strains in *Drosophila melanogaster* and *D. simulans*. The differential effects of temperature selection on life-cycle components of a spectrum of genotypes could greatly increase the capacity of the population for adaptation in a heterogeneous and ever-changing environment. At the ecobehavioral level, the importance of environment far transcends its relative importance in the study of phenotypes at the morphological level. However, once metabolic processes are involved in any way, then temperature is a variable that must be considered for a trait. In *Drosophila*, the complications due to the environment are not serious, since genotype–environment interactions can be estimated by appropriate statistical designs, as can the main effects of genotypes and environments. In rodents, superimposed upon direct environmental effects are possible complications due to learning. Consequently, the study of the ecobehavioral phenotype becomes progressively more difficult moving up the phylogenetic series, culminating in our own species for which environments cannot be controlled as in experimental organisms where breeding experiments can be done.

In this book, the ecobehavioral phenotype has been looked at in relation to four major components of the environment – weather, resources, other organisms, and habitats. Although these categories overlap to some extent, they are sufficiently discrete to be considered separately in the main. Interestingly, these are the major categories considered by Andrewartha and Birch (1954) in their classic *The Distribution and Abundance of Animals*, and they represent selection regimes (Wallace, 1981) ranging from hard to relatively soft. Experimental results encompassing these environmental components are usually quantitative rather than qualitative. Hints as to the genetic architectures of ecobehavioral phenotypes are obtainable in genetically well-known organisms, such as *Drosophila melanogaster*, and suggest that rather few genes account for most of the variability. Indeed, the situation for many traits appears to be a few locatable genes having relatively large effects modified by many of small effect all interacting in various ways to produce almost universal pleiotropy. The discussions in Chapters 5 and 6 indicate that the relationship of these genes with electrophoretic and chromosome polymorphisms studied so extensively in natural populations is at the present moment rather obscure.

This means that at the natural population level, it is necessary to turn to the methods of quantitative genetics. Studies on natural populations are now increasing in number. It is conventionally assumed that differences among geographical populations imply natural selection in relation to diverse habitats. Accordingly, it may be important to estimate heritabilities and genetic correlations among traits of eco-behavioral significance. For example, the feeding responses of discrete races of newborn garter snakes (Section 8.3) show little or no geographical variation for estimates of such parameters. Because the widespread application of quantitative genetic theory to evolution in nature would be immeasurably assisted if it could be assumed that such estimates do not vary much for various suites of traits, the importance of additional studies is obvious. However, for ecological phenotypes, major differences between populations for isofemale heritabilities from benign and stressful habitats (Section 4.3) implies that this assumption will often be an oversimplification. Generalizations may, therefore, be difficult to obtain and should be approached cautiously.

In this book, the $r–K$ continuum of selective characteristics is used in discussions of life histories merely for simplicity. It is not the function of this book to discuss the relative merits of the alternative approaches to the classification of life histories. In r-selected species, such as *Drosophila melanogaster* and *D. simulans*, little variation in development time is expected, because simplistically this trait should be under continual selection for rapid development. Yet within and among populations of the same species there are exceptions. In climatically benign habitats, development in *D. melanogaster* is slower and more variable than in climatically stressful habitats (Section 4.3). Combined with evidence in Sections 2.1, 7.2, and 8.2, development time is more variable than the $r–K$ continuum model predicts. It is not, therefore, surprising that there is some genetic variability for the trait in natural populations, in agreement with recent evidence and theoretical arguments (Rose and Charlesworth, 1981) that significant additive components of variance are not unusual for fitness traits in outbred populations. It appears likely that fitness traits vary somewhat depending upon habitat, but not to the extent that traits more peripheral to fitness may vary among populations for which stabilizing selection will be important. Population differentiation mainly at the geographical level within species implies past directional selection perhaps occurring sporadically, so that both additive and dominance variation may be usual for fitness traits.

Traits are continually being found that deviate from the complex packages of life-history traits predicted under the $r–K$ continuum and alternative life-history classifications. For example, Atkinson (1979) showed that the larval niches of *Drosophila* must be a major factor in establishing the diversity of female reproductive systems among *Drosophila* species (Section 9.1) and a suite of reproductive traits shows inconsistencies with the $r-K$ continuum predictions. In addition, there is no reason why only one system of life-history arrangements among traits is possible in a particular environment. It may be more reasonable to assume that fitness at the organismic level is maximized in the same environment in a number of different ways. In *D. melanogaster*, the isofemale strain technique will be useful in investigating the possibility (or otherwise) of trade-offs among life-history traits, as will quantitative genetic studies to estimate their components of variance. Comparisons of colonizing species with closely related noncolonists in this way may provide an underlying understanding of the ecobehavioral phenotypes important in colonization phenomena.

12.2 Colonists and habitats

Based firmly upon the organism as the unit of selection, the aim here is to consider briefly two of the important questions that occur when thinking about colonists and their habitats. They can only be partially answered at this stage.

    1. *Are range expansions ultimately prevented by a lack of genetic variability for further adaptation to novel habitats?*

Summarizing some of the discussions in this book, an ecological range expansion into a new habitat implies directional selection for adaptation to the environmental components of that habitat and a response by the colonizing population. The response must be sufficient that some individuals survive through the population crashes of stressful environments. At the genetic level, high flexibility is obtained via a genome where recombination is not unduly restricted by inversion polymorphisms. In this way, genes may be rapidly combined together so that in theory a phenotype that is well adapted to the new habitat may evolve without difficulty. Provided that genetic variability is present, then such adaptation could occur over a relatively short time period, assuming that genes controlling ecobehavioral phenotypes have relatively large effects. However, any new and extreme genetic combination is likely to have a lower overall fitness as measured by viability, fecundity, and, hence, r. This may limit colonization potential, just as many laboratory directional selection programs have been abandoned when fitness falls to unmanageably low levels.

In assessing the possibility of range expansions, it is important to consider population size–variability relationships. Heterozygosity is not reduced to impossibly low levels for a founder population size of

Table 12.1. *Mean scutellar bristle numbers in 16 isofemale strains of* Drosophila melanogaster *from Leslie Manor, Victoria*

Strain	Mean	Strain	Mean
1	4.03	25	4.02
2	4.18	26	4.06
3	4.00	27	4.04
20	4.05	29	4.01
21	4.00	31	4.02
22	4.02	32	4.01
23	4.03	33	4.02
24	4.10	34	4.01

Source: After Hosgood, MacBean, and Parsons (1968).

two, especially if r is reasonably high as is likely in colonizing populations (Section 4.4). Such a population size represents an isofemale strain without the complication of multiple insemination. It has been shown that isofemale strains vary among themselves by chance and this variation may persist for many generations. Although this applies to ecobehavioral traits, such as tolerance to temperature extremes and mating behavior traits, most data have been collected on bristle number at various sites on adults of *Drosophila melanogaster* (Parsons, 1980a). The persistence of variation among isofemale strains for many generations suggests that a method of obtaining rapid responses to directional selection is to base selection upon an extreme strain, or a combination of extreme strains, from a set of isofemale strains.

The number of scutellar bristles on most wild flies of *D. melanogaster* is four, two on each side; however, a small number of flies in populations have five and even six bristles. Certain isofemale strains have considerable proportions of flies with more than four bristles persisting at characteristic proportions for many generations. This variation among strains derives from the founder females from the wild population that is polymorphic for bristle genes (Section 4.4). Sixteen isofemale strains were set up at random from a wild population with mean bristle numbers given in Table 12.1 (Hosgood and Parsons, 1967). Seven generations of directional selection were carried out (Figure 12.1) based upon various combinations of isofemale strains:

1. The highest isofemale strain – an initial population size of N = 2
2. A hybrid of the highest four isofemale strains – an initial population size of $N = 8$

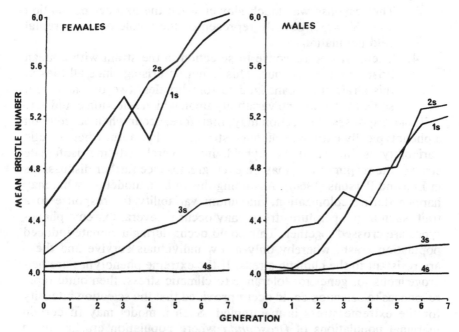

Figure 12.1. The mean scutellar bristle number in four selection lines 1 to 4 in *Drosophila melanogaster* as described in the text. (Source: Simplified from Hosgood and Parsons, 1967.)

3. A hybrid of all 16 strains – an initial population size of $N = 32$.

4. The isofemale strain with the lowest mean – an initial population size of $N = 2$.

The results in Figure 12.1 show that

1. Small founder population sizes ($N = 2, 8$) of isofemale strains with high bristle numbers are conducive to rapid responses to selection. These responses are caused by the accumulation of bristle genes with reasonably large effects, as has been demonstrated by the genetic location of some of these genes (MacBean, McKenzie, and Parsons, 1971). A genetic architecture of many genes of small effect is inconsistent with such rapid responses.

2. Although the initial bristle number from the larger of these two founder populations ($N = 8$) was lower than that from the smaller ($N = 2$), the ultimate response to selection was greater, presumably because of a greater diversity of high bristle number genes in the former situation. That is, selection among isofemale strains in this way may bring together a greater array of high bristle number genotypes for $N = 8$ than for $N = 2$.

3. The response was much slower when the founder population size (N = 32) directly represented the whole of the parental wild population.
4. There was no response to selection in the strain with a mean bristle number of four. This is not surprising since all flies in this strain were canalized to four bristles, two on each side, so that there was no variability upon which selection could act.

Rapid responses to selection may, therefore, occur when based upon a phenotypically extreme isofemale strain. Furthermore, when genetic variability is increased by hybridizing several extreme isofemale strains, the response rate may be even greater [see further discussions in Lee and Parsons (1968)]. Assuming this to be a model for what may happen during colonization, maximum variability for response to a trait, such as temperature stress, may occur if several extreme phenotypes are crossed together. This could occur during a climate-induced population crash, whereby only a few individuals survive and these are resistant to the climatic stress. If the extreme phenotypes are heterogeneous for genes for tolerance to climatic stress, then quite rapid responses to selection could occur provided that the selection intensity for the extreme stress is maintained. Such a model may fit certain marginal populations of *Drosophila* where population crashes often occur. If the selection for the extreme stress is not maintained, such genes will be rapidly diluted during the r growth phase following the population crash. One caveat is that extreme populations may have low variability for ecological phenotypes under direct natural selection (Section 4.3), and this may be rapidly restrictive. It is important to reiterate, however, that these populations will not normally be expected to have parallel reduced variability for electrophoretic variants. That is, strong selection at the ecological phenotypic level may be translated into weak to nonexistent selection at the protein/enzyme polymorphism level.

There is always the possibility that populations will be totally eliminated during a crash, that is, via hard selection. This may be quite usual, but if combinations of genes conducive to stress-resistant phenotypes occur even if rarely, stable range expansions may ensue on an occasional basis. It, therefore, may be important for populations to go through fairly frequent crashes so that genes controlling extreme phenotypes permitting range expansions are favored. Migration across habitats could increase the chances of diverse genes from different habitats favoring range expansions being combined together. Given the distances certain colonizing insect species can move over a short time period (Section 9.4), this could be more common than would have been envisaged a few years ago. This appears also to be true of plants. In the sweet vernal grass, *Anthoxanthum odoratum*, Grant and Antonovics (1978) record that although the proportion of pollen originating from central populations that reached marginal populations was small, the relatively much larger sizes of the central populations resulted in substantial potential cross contamination. Accidents of sampling will, therefore, play a major role in the determination of the genotypes of the successful migrants, and more generally, which genotypes survive in the density-independent conditions occurring in ecologically marginal habitats compared with more benign habitats. Ultimately, a lack of genetic variability or poor fitness of extreme phenotypes will prevent range expansions. Beyond the ecobehavioral margin so formed, selection would be completely hard.

2. *Can phenotypic (and genotypic) criteria be established to determine those species likely to be colonists?*

This question can be attacked at the phenotypic level by considering Andrewartha and Birch's (1954) four major components of the environment (Section 12.1). A range expansion will involve an ecobehavioral change in one or more of these components. Unfortunately, range expansions are rarely witnessed, although the increased tolerance to temperature extremes in *Dacus tryoni* as it spread south in Australia is one example. Considering *Drosophila*, the plot of species in Figure 3.3 shows that colonizing species must evolve a certain minimal level of tolerance to climatic extremes. Comparisons of colonizing species with closely related noncolonists for responses to directional selection in experiments of the type described earlier for scutellar bristle number would be useful for the assessment of colonization potential at both the phenotypic and genetic levels. This is because of the need to adapt to ecologically marginal habitats, where physical stresses of climatic origin tend to be variable and extreme. The genes underlying responses to selection can then in theory be localized and studied individually. From such analyses, it will be possible to see the extent to which easily detectable gene and chromosome polymorphisms contribute to largely ecobehavioral phenotypes. In addition, using isofemale strains direct from natural populations, correlations among traits can be calculated, to assess the degree of pleiotropy versus linkage among traits. This may be a useful approach in the study of the relative importance of the various life-history traits that are of relevance in determining the potential for colonization, or the study of resistance to environmental stresses, say cold or heat stresses, at different development stages.

Potential colonists (in *Drosophila*) need to be close enough to the temperate-zone plot in Figure 3.3 to have variation permissive of entry into this sector. On this criterion, the Australian endemic species *D. inornata* is a potential colonist, but this species is restricted by being genetically alcohol dehydrogenase (*Adh*)-null, so is unable to exploit the resources of disturbed habitats as can species with active *Adh* alleles. However, in spring when *D. inornata* is found in urban habitats (Section 3.3), an active *Adh* allele could be favored, so that colonization could theoretically follow if such a mutant occurred. This type of change is supposed to have occurred in the evolution of the genus *Drosophila* when adapting to fermented-fruit resources (Throckmorton, 1975).

In conclusion, interspecific comparisons for Andrewartha and Birch's (1954) components of the environment, especially for phenotypes related to weather and resources, help to establish criteria for colonization potential. In ecologically marginal habitats, Andrewartha and Birch's two remaining components – other organisms and habitats – will be less important; competition is minimal in *r*-selected species, and precise adaptation to specific habitats is less likely than when conditions are more benign. As yet, work on the requisite genetic analyses has hardly begun.

12.3   Discontinuities in evolution

Darwin (1859) argued that the origin of species by natural selection proceeded by the accumulation of slight successive favorable variations, producing no great or sudden morphological shifts. In contrast, a number of evolutionary biologists and paleontologists (e.g. Simpson, 1953; Grant, 1963; Mayr, 1963; Eldredge and Gould, 1972; Stanley, 1979) have argued that rapid morphological change may accompany speciation, for which Grant (1963) proposed the term *quantum speciation*. Stanley (1979), in particular, has developed these ideas at the macroevolutionary level. A recent example of the type of data being accumulated comes from Williamson (1981), who studied a sequence of fossilized freshwater mollusks in East Africa, which have remained more or less undisturbed since their formation. The order of formation of each bed can be identified by referring to their relationship with accurately dated geological features. In 13 lineages common enough for detailed analysis, long periods of morphological stability are interrupted by fossil beds in which relatively rapid changes in shell shape take place. At this time there may have been synchronous environmental stresses affecting all lineages in parallel ways. The intermediate forms between the ancestral and derived species occupy a very small proportion of the evolutionary history of each lineage. Additionally, the periods of transition coincide with each other in the various genera, perhaps corresponding to periods of ecological change. As Jones (1981) points out, this argues that "evolution is seen as an essentially discontinuous process in which long periods of genetic stability are interrupted by 'genetic revolutions' which produce very rapid change,

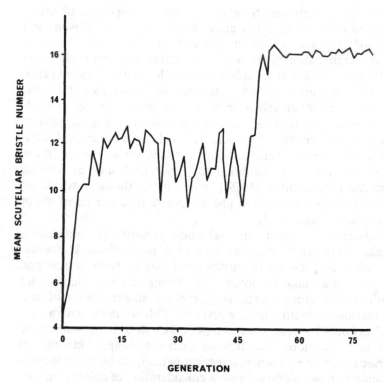

Figure 12.2. The ultimate response to selection in females for scutellar bristle number in line 2 of Figure 12.1. Note the two accelerated responses lasting very few generations in each case. (Source: Simplified from MacBean, McKenzie, and Parsons, 1971.)

rather than as a system of gradual change leading to the origin of species acting on minor differences in fitness among individuals.''

The intermediate forms in Williamson's fossil series existed for between 5000 and 50,000 years, which are periods much shorter than those of evolutionary stability of each lineage but may still take of the order of 20,000 generations to complete. This is a long time in the context of artificial laboratory selection experiments. In many such experiments, a process of gradual response to selection without marked discontinuities occurs as in Figure 12.1 over seven generations. However, plotting females of strain 2 in Figure 12.1 over a period of 75 generations reveals that the major response to selection for scutellar bristles occurred in two rapid phases, or accelerated responses to selection, lasting a few generations in each case (Figure 12.2). Indeed, the increase from 10 to about 16 bristles occurred over four generations. The time period during the plateau phases vastly exceeded that of the accelerated response phases. Similar accelerated responses to selection have been reported by other authors (e.g., Thoday, Gibson, and Spickett, 1964; and see review in Lee and Parsons, 1968). In some cases, gene location studies show that recombination between genes of relatively major effect can explain the accelerated responses (Thoday, 1961; Lee and Parsons, 1968; MacBean, McKenzie, and Parsons, 1971). Therefore, although selection responses may often appear gradualistic, accelerated responses interrupting plateaus are analogous to periods of stability interrupted by genetic revolutions on an evolutionary time scale. In the Antarctic radiolarian, *Pseudocubus vema*, there appears to be an analogous situation to Figure 12.2 [data of Kellogg (1975) discussed by Stanley (1979)], since mean thoracic width increased in two major accelerated phases over a time period of about 2.5 million years (Figure 12.3).

In the laboratory experiments, selection pressure is continuous, which explains quite rapid changes over a few generations. However, in nature, selection pressure is variable and may be relaxed now and then, so that it may take far longer for change to occur than under highly artificial laboratory situations. Therefore, an experiment of continuous directional selection lasting just over 75 laboratory generations (Figure 12.2) could be analogous to a period of an order of magnitude greater in nature or more. This follows from some unpredictability of stress periods (Figure 6.1) when selection is likely to be most intense for adaptation to new habitats. From a consideration of colonizing species, it appears that major evolutionary transitions would tend to occur during times of overall environmental change in a given direction. Hence, genetic changes in organisms during a colonization phase at the intraspecific level appear no different in principle to changes occurring on an evolutionary time scale of the type studied by Stanley (1979) and Williamson (1981). In both cases, changes may occur over short time intervals in the total time interval under study so that discontinuities prevail. At the intraspecific level, the discontinuities may arise from the underlying genetic architectures of quantitative traits. During speciation, this underlying genetic architecture of a few major additive genes largely controlling quantitative traits may lead to both pre- and postmating isolation from the ancestral population, as well as differences in morphology, life history, development, physiology, and so forth. (Templeton, 1980a, 1981); ecobehavioral traits may be added to this list (Parsons, 1982a).

12.4 Future directions

Predicting future trends in any area of science is tricky; nevertheless it should be attempted. Among important topics for future study are

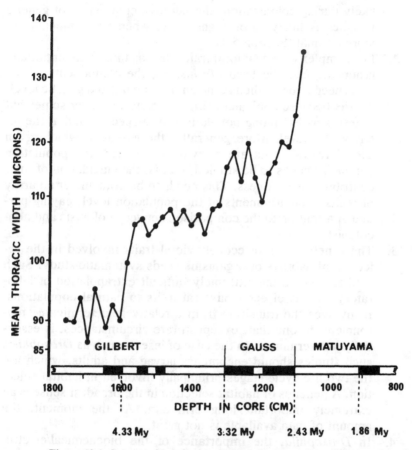

Figure 12.3. Increase in thoracic width of the Antarctic radiolarian, *Pseudocubus vema*, during an interval of about 2.5 million years. Alternating black and white bars depict reversals of the earth's magnetic field that have been measured in deep-sea cores from which samples were obtained. (Source: Simplified from Kellogg, 1975, and Stanley, 1979.)

1. Many of the laboratory studies are based upon a restricted set of genotypes in a restricted set of environments. Calculated heritabilities are frequently based upon studies on one population in a single environment. When the number of genotypes and environments affecting an ecobehavioral trait is considered the problem of generalization is enormous. It becomes incredibly complex if the environment is defined in the broadest sense to include not only the physical environment, but also the effects of previous experience. Future endeavors will have to come to grips with these complexities. From the analytical point of view, it is convenient that under the extreme stresses

likely during colonization, the additive component of genetic variance is likely to be higher than when environments are more optimal (Section 5.4).

2. For complex ecobehavioral traits such as those conventionally grouped under the term *life history*, the components of the traits need to be studied at the environmental and genetic level. This is because equivalent fitnesses may occur by somewhat differing paths among populations of a species, and at the interspecific level. More generally, the extent to which given ecobehavioral traits are relevant in determining population continuity needs to be studied, that is, the elucidation of their contributions to fitness. This needs to be done under an array of realistic environments at the population level, paying particular attention to the comparison between colonists and noncolonists.

3. The genetic basis of ecobehavioral traits involved in the selection of habitats of organisms needs systematic study. Such studies involve the extremely difficult extrapolation of laboratory studies of ecobehavioral traits to natural populations. It involves the transition from a relatively determinate environment, to one that, except in rare circumstances, is essentially indeterminate. In the case of insects such as *Drosophila*, such studies should encompass larvae and adults, which are the two life-cycle stages principally involved in habitat selection. A genetics of habitat selection in the broadest sense is an extremely important long-term aim. At the moment, the amount of data available is not great.

4. In *Drosophila*, the importance of the biochemical/electrophoretic phenotype in indicating colonizing potential was shown with reference to phenotypes at the *Adh* locus, ADH enzyme activity, and ethanol resource utilization. Can biochemical phenotypes be established in other colonizing species that aid in comparisons with closely related noncolonists?

5. It is important to study actual colonizations following artificial or other major environmental changes, including introductions of organisms for the purposes of biological control. During the introduction, simultaneous information on ecobehavioral and genetic variables is needed. It will be necessary to investigate effective population size especially at the bottleneck, migration rates after introduction, gene frequencies from which genetic distances can be obtained, ecobehavioral phenotypes and their underlying genotypes, and life histories. Information along these lines is needed during the whole colonization episode to

provide a comparison of the introduced and source populations.

6. Migration rates are difficult to study. There is suggestive evidence that adults of certain species of *Drosophila* and the tephritid fly, *Dacus*, can migrate substantial distances. Comparative studies of migration rates of colonizing species with noncolonists are needed. One prediction is an association of migration with high resistance to physical stresses, so that there may be success for colonizing voyages in migratory species.

7. In both of the applied areas discussed briefly – biological control and domestication – the integration of the study of behavior, ecology, and genetics is a prerequisite for predictive purposes. The complexity of this task becomes daunting, when it is realized that this encompasses all of the studies in topics 1 to 6!

8. Although not discussed in detail, the array of reproductive systems found especially in parasites and plants needs to be investigated more fully with colonization potentials in mind.

9. The final point is one that has been constantly discussed – namely the relationship between the ecobehavioral phenotype of an organism and its underlying genotype. Predictions on the possibility of success of colonization, whether naturally or for biological control, depend upon a full understanding of interrelationships between the organizational levels of genotype, organism, and population.

Appendix
The study of quantitative traits

> The biometrical facts as to the inheritance of stature and other human measurements, though at first regarded as incompatible with the Mendelian system, have since been shown to be in complete accordance with it, and to reveal features not easily explicable on any other view. [Fisher, 1930: 17]

A.1 Variability

Various concepts from the field of quantitative genetics have been used in this book. They are developed far more fully in Kempthorne (1957) and Falconer (1960). A less rigorous presentation, but useful in the context of this book since ecobehavioral examples are used, appears in Ehrman and Parsons (1981a) with earlier accounts in Parsons (1967, 1973b). However, in the interests of completeness, a brief survey is presented here.

The frequency distribution of many quantitative traits approximate, more or less, the continuous normal distribution of the statistician (Figure A.1). This distribution can be completely described in terms of two parameters, based upon individual observations, x_i:

$$Mean\ \bar{x} = \Sigma x_i / n$$

where there are n observations, and

$$Variance\ V(x) = \frac{1}{n - 1} \Sigma (x_i - \bar{x})^2$$

which provides an expression of variability around the mean.

Assuming that a continuously varying trait is partly under genetic control, how is the intrinsically discontinuous variation caused by genetic segregation expressed as continuous variation? Suppose two individuals $A/a.B/b$ are crossed where A,a and B,b are gene pairs at two unlinked loci, and where genes A and B act to increase the measurement of a quantitative trait by one unit, and genes a and b act to decrease the trait by one unit. Rewriting $A/a.B/b$ as $+/-.+/-$, where A and B are $+$ genes, and a and b are $-$ genes, a metric or quantitative value

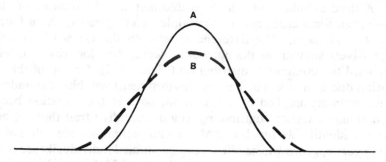

Figure A.1. Normal distribution. Curves A and B have the same mean, but the variance of B exceeds that of A.

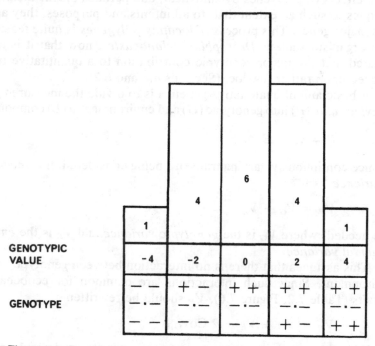

Figure A.2. Genotypic frequencies from the cross $+/-.+/- \times +/-.+/-$ plotted according to the genotypic value (i.e., the relative number of + and − genes). The frequencies of each genotype are given in the histogram.

of a genotype can be obtained by counting the number of + and − genes. This cross gives five genotypes (Figure A.2), ranging from one with four − genes to one with four + genes. The most common genotype is $+/-.+/-$ having a *genotypic value* of zero, since it has two + and two − genes. This is the mean genotypic value.

A third unlinked $+/-$ locus would increase the number of classes to seven for a cross between multiple heterozygotes, and a fourth to nine, and so on. The differences between the classes become progressively smaller, as the number of segregating loci rises, so tending toward the continuous distribution (Figure A.1). On top of this, variation due to nongenetic causes (environment) will blur the underlying discontinuity implied by segregation, so that the variation becomes continuous. Genes contributing to a quantitative trait that are not directly identifiable by classic Mendelian techniques are referred to as *polygenes*; genes whose effects can be studied individually are referred to as *major genes*. There is no fundamental biological distinction between the two categories; the two terms are merely a matter of convenience. Indeed, it is possible in certain circumstances to increase the effects of polygenes by statistical, and perhaps biochemical, techniques to such an extent that, to all intents and purposes, they appear as major genes. This process of *locating polygenes* is quite feasible in an organism such as *Drosophila melanogaster*, now that it is appreciated that the major genotypic contribution to a quantitative trait is based upon rather few loci (Sections 4.4 and 6.2).

A basic aim of quantitative genetics is to divide the measured *phenotypic value* (P) into genotypic (G) and environmental (E) components:

$$P = G + E$$

Since continuously varying traits are being considered, the *phenotypic variance*

$$V_P = V_G + V_E$$

is needed, where V_G is the *genotypic variance* and V_E is the *environmental variance*.

This assumes that there is no interaction between genotype and environment. Since such interactions are common for ecobehavioral traits (Table 7.2, Figure 7.6), V_P should be rewritten

$$V_P = V_G + V_E + 2\,\mathrm{Cov}_{G,E}$$

where $\mathrm{Cov}_{G,E}$ represents the covariance of genotypic and environmental values. The presence of *genotype–environment (GE) interactions* will increase V_P. This is not surprising, since these interactions increase the level of observed variability.

Such GE interactions must be distinguished from those where there is a *correlation between genotype and environment*. If there is a positive GE correlation, V_G is overestimated, and if negative, underestimated. In human beings, it has been suggested that in a favorable environment, genetic effects might be expressed more fully than otherwise; indeed, positive correlations between genotype and environment frequently occur in human data on intelligence (Ehrman and Parsons, 1981a). An agricultural example is where the better animals are given more food. In natural populations, animals may seek the microhabitat for which they are best suited, that is, habitat selection (see Chapters 8 and 9) may lead to GE correlations.

More often than not, GE interactions are ignored. In these circumstances, there are simple experimental techniques available for computing the proportion of the total variance that is genotypic, or

$$\frac{V_G}{V_G + V_E} = \frac{V_G}{V_P} = h_B^2$$

a ratio called *the heritability in the broad sense* or *the degree of genetic determination of a trait*. Clearly $0 \leq h_B^2 \leq 1$, ranging from a trait determined entirely environmentally to a trait where the correspondence between genotype and phenotype is direct, without any environmental influence whatsoever. Measures of laterality form good examples of traits with heritabilities not usually differing significantly from 0 (Ehrman and Parsons, 1981a). Quantitative traits normally fall between the extremes of 0 and 1.

Now, considering two alleles at a locus A_1 and A_2, there are three possible genotypes – A_1A_1, A_1A_2, and A_2A_2 – two being homozygotes and one a heterozygote. If the genotypic value of the heterozygote is equal to the mean of two homozygotes, then there is no dominance. Assuming the presence of dominance and epistasis (i.e., interactions between loci), the genotypic variance can be subdivided

$$V_G = V_A + V_D + V_I$$

where V_A = variance due to additive genes (*additive genetic variance*), V_D = variance due to dominance deviation (*dominance variance*), and V_I = variance due to epistasis (*epistatic variance*). Where estimated, V_I is usually relatively small, so the phenotypic variance can be approximated as

$$V_P = V_A + V_D + V_E.$$

The ratio V_A/V_P is the *heritability in the narrow sense*, h_N^2, and is a measure of variation due to additive genes. It is a more useful concept than h_B^2 defined earlier, because when considering relationships between generations, it is the gametes carrying genes rather than genotypes that are passed on from one generation to the next. From the predictive point of view, therefore, estimates of h_N^2 should be obtained where possible.

A.2   Estimating components of variability

There are various methods for estimating heritability and associated parameters. Most of the following have been mentioned, directly or indirectly, with reference to data in this book.

Variability within and between inbred strains

Inbred strains maintained by brother–sister (sib) mating have been developed in many species, in particular, rodents and *Drosophila*. In theory, they are homozygous so that all variation within strains is environmental, and variation between strains comprises both genotypic and environmental components. Based upon an analysis of variance, if the mean square between strains is M_1 and that within strains is M_2, and r is the number of replicates per strain, it can be shown that

$$V_G = \frac{M_1 - M_2}{r} \quad \text{and} \quad V_E = M_2$$

from which h_B^2 can be estimated. By carrying out such an analysis over a set of environments, say temperatures and light intensities, a comprehensive assessment of various GE interactions becomes additionally possible.

Crossing two inbred strains and derivative generations

By taking two homozygous inbred strains P_1 and P_2, and various crosses to give the F_1, F_2, BC_1, and BC_2 (backcrosses of F_1 to P_1 and P_2, respectively) generations, as well as all the possible reciprocal crosses, a total of 14 crosses can be set up. This permits the separation of the genotypic variance into additive, dominance, epistatic, and reciprocal (i.e., when there are differences according to which parental phenotype is male and which is female) components.

Diallel cross

A *diallel cross* is the set of all possible matings between several strains or genotypes. For four strains, there are 16 combinations made up of six crosses AB, AC, AD, BC, BD, and CD, their six reciprocals where the sex of the parents is transposed, and four kinds of offspring derived from the four parental strains, AA, BB, CC, and DD, which are arranged along the leading diagonal (Table A.1). Diallel crossing techniques vary according to whether the parental strains or the reciprocals are included, or both. There are a variety of theoretical methods of analysis of diallel crosses available depending somewhat upon the in-

Table A.1. *Four-strain diallel cross*

Strain of female parent	Strain of male parent			
	A	B	C	D
A	AA	AB	AC	AD
B	BA	BB	BC	BD
C	CA	CB	CC	CD
D	DA	DB	DC	DD

formation desired (Griffing, 1956; Kempthorne, 1957; Mather and Jinks, 1977). To obtain estimates of V_A and V_D, and, hence, h_N^2 and h_B^2, the strains A, B, C, and D are assumed to be inbred.

Now the average of a given strain in hybrid combination is referred to as the *general combining ability* (gca) of that strain. Normally gca's are expressed as deviations from the overall population mean. In the aforementioned diallel, gca's for strains A, B, C, and D can, therefore, be computed (Griffing, 1956).

Given the gca's of two strains, say A and B, it is possible to predict the value of crosses between them, that is, for A ♂ × B ♀ and its reciprocal. The degree to which the mean of this specific cross differs from that predicted on the basis of the gca's of strains A and B is referred to as the *specific combining ability* (sca). Finally, it is possible to calculate the degree to which A ♂ × B ♀ differs from its reciprocal B ♂ × A ♀, which is referred to as a *reciprocal effect* (r).

From an analysis of variance, it is possible to estimate V_{gca}, V_{sca}, V_r, and V_E. Clearly, V_{gca} must have a major additive genetic component, and V_{sca} a major dominance component. If A, B, C, and D are inbred strains, it turns out that (Griffing, 1956) $V_A = \frac{1}{2}V_{gca}$ and $V_D = V_{sca}$, from which h_N^2 and h_B^2 can be calculated.

Relationships between relatives

Since inbred strains are not available in all organisms, especially vertebrates and humans, *relationships between relatives* are commonly studied. Especially common is the method of investigating correlations between parents and their offspring and between sibs. This is becoming an increasingly sophisticated area, involving not just one trait, but also groups of traits involved in resource-utilization and life-history studies. Table A.2 gives expected covariances between relatives and the correlation coefficients so obtained expressed in terms of h_N^2. (Note: In the case of full sibs including twins, the common parents mean that

Table A.2. *Correlations between relatives*

Relatives	Covariance between relatives	Correlation coefficient
Monozygotic twin	$V_A + V_D$	$> h_N^2$
Midparent, child	$\frac{1}{2}V_A$	$h_N^2/2^{1/2}$
Parent, child	$\frac{1}{2}V_A$	$\frac{1}{2}h_N^2$
Sib, sib ⎱ Dizygotic ⎰	$\frac{1}{2}V_A + \frac{1}{4}V_D$	$> \frac{1}{2}h_N^2$
Half sibs ⎫ Uncle, nephew ⎬ Aunt, niece ⎭	$\frac{1}{4}V_A$	$\frac{1}{4}h_N^2$
First cousin	$\frac{1}{8}V_A$	$\frac{1}{8}h_N^2$
Second cousin	$\frac{1}{32}V_A$	$\frac{1}{32}h_N^2$

some genotypes will be in common so a dominance component appears in the covariance, thus, the correlation coefficient overestimates h_N^2.)

Relationships between relatives can also be investigated using a regression approach. If b_{OP} is the regression of one offspring on parent, then it can be shown that

$$b_{OP} = \frac{1}{2}h_N^2$$

and for the regression of offspring on midparent $\bar{P}$,

$$b_{OP} = h_N^2$$

Directional selection

The selection experiment consists of selecting and manipulating various chosen phenotypes with respect to a trait from a population. The concern here is *directional selection* (Figure A.3), where extreme individuals from a population are selected in the hope of forming separate high and low lines in subsequent generations. If a quantitative trait has some genetic basis, there should be a response to directional selection since selection of extreme phenotypes means that extreme genotypes are selected. The *response to selection R* can be expressed as

$$R = b_{OP}S,$$

where b_{OP} is the regression of offspring on midparent, and S is the *selection differential* (Figure A.3). It follows that

$$R = h_N^2 S \quad \text{or} \quad h_N^2 = R/S$$

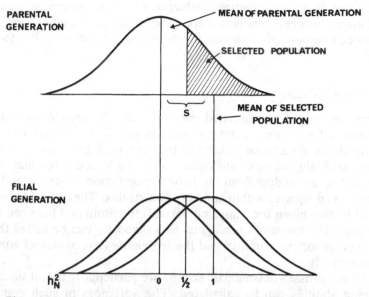

Figure A.3. Diagram showing the meaning of the selection differential, S, where all individuals in the shaded region are selected in the parental generation. The response to selection, R, depends upon the heritability in the narrow sense, h_N^2, as shown in the filial generation curves.

which is referred to as the *realized heritability* and calculated over a number of generations of selection. As can be seen in Figure A.3, the selection differential, S, is defined as the mean phenotypic value of individuals selected as parents, expressed as a deviation from the mean phenotypic value of all individuals in the parental generation before selection was commenced. Clearly, the greater the response to selection, R, the higher the value of h_N^2. An example of directional selection for cold resistance in *Drosophila melanogaster* is given in Section 4.2.

Correlated response

It is often useful to study correlated responses if several traits are being studied, in addition to the trait under direct selection. Correlated responses to selection may be positive or negative. In genetic studies, it is necessary to distinguish the causes of correlation between traits – genetic and environmental. The directly observable correlation between two phenotypic values for traits x and y is the *phenotypic correlation* (r_P). Similarly r_A is the *genetic correlation* between the additive genetic values of x and y, and r_E the *environmental correlation* between the traits. To compute genetic correlations, relationships between relatives or data from directional selection experiments can be used. Correlations between relatives were used to obtain genetic correlations between responses to various forms of prey in garter snakes (Section 8.3).

Isofemale strains

Some of the aforementioned methods can be applied to isofemale strains set up from natural populations (Sections 4.3 and 4.4). From an analysis of variance within and between isofemale strains, it is possible to obtain quantities analogous to V_G and V_E if the isofemale strains are set up at random from the natural population. These would come from mean squares within and between strains. The procedure is identical to that given for analyzing variability within and between inbred strains. The quantities analogous to V_G and V_E can be called the *isofemale genotypic variance* and the *isofemale environmental variance*, respectively.

Diallel crosses of isofemale strains are particularly useful since combining abilities can be calculated. The variances of such combining abilities V_{gca} and V_{sca} will not provide direct estimates of V_A and V_D since inbred strains must be used for this purpose. However, values of V_{gca} and V_{sca} do provide an excellent idea as to the relative importance of additive and dominance genes for a set of traits in a given population. Indeed, from the diallels of isofemale strains can be obtained estimates that can be called the *isofemale additive genetic variance* and *isofemale dominance variance*.

From these variance estimates can be obtained *isofemale heritabilities,* analogous to h_B^2 and h_N^2. In using this terminology, it is important to stress the analogy with the populational heritability of Slatkin (1981), which is derived from comparisons within and among populations (see Section 7.3) of varying sizes. Some estimates of isofemale heritabilities analogous to h_B^2 are presented in Tables 4.2, 4.3, and 5.5 based on analyses of variance within and between isofemale strains. As stressed earlier, provided that the isofemale strains are set up at random from natural populations, all of these estimates will be applicable directly to the population itself.

For rapid responses to directional selection, therefore, the isofemale additive genetic variance and, hence, the corresponding isofemale heritability should be high. Isofemale strains have the additional advantage in that directional selection itself can be based upon an extreme strain or a combination of extreme strains from a set of isofemale strains (Section 12.2).

References

Aceves-Piña, E. O., and Quinn, W. G. 1979. Learning in normal and mutant *Drosophila* larvae. *Science 206*:93–6.

Allard, R. W. 1965. Genetic systems associated with colonizing ability in predominantly self-pollinated species. In *The Genetics of Colonizing Species* (edited by H. G. Baker and G. L. Stebbins), pp. 49–76. New York: Academic Press.

Allemand, R., and David, J. R. 1976. The circadian rhythm of oviposition in *Drosophila melanogaster*: a genetic latitudinal cline in wild populations. *Experientia 32*:1403–5.

Ammerman, A. J., and Cavalli-Sforza, L. L. 1971. Measuring the rate of spread of early farming in Europe. *Man 6*:674–88.

Anderson, W. W. 1968. Further evidence for co-adaptation in crosses between geographic populations of *Drosophila pseudoobscura*. *Genet. Res. 12*:317–30.

– 1974. Frequent multiple insemination in a natural population of *Drosophila pseudoobscura*. *Am. Nat. 108*:709–11.

Anderson, W. W., Levine, L., Olvera, O., Powell, J. R., de la Rosa, M. E., Salceda, V. M., Gaso, M. I., and Guzmañ, J. 1979. Evidence for selection by male mating success in natural populations of *Drosophila pseudoobscura*. *Proc. Natl. Acad. Sci. USA 76*:1519–23.

Andrewartha, H. G., and Birch, L. C. 1954. *The Distribution and Abundance of Animals*. Chicago: University of Chicago Press.

Anxolabehere, D., and Periquet, G. 1970. Résistance des imagos aux basses températures chez *Drosophila melanogaster*. *Bull. Soc. Zool. Fr. 95*:61–70.

Arnason, E., and Grant, P. R. 1976. Climatic selection in *Cepaea hortensis* at the northern limit of its range in Iceland. *Evolution 30*:499–508.

Arnold, S. J. 1981a. Behavioral variation in natural populations. II. The inheritance of a feeding response in crosses between geographic races of the garter snake, *Thamnophis elegans*. *Evolution 35*:510–15.

– 1981b. Behavioral variation in natural populations. I. Phenotypic, genetic and environmental correlations between chemoreceptive responses to prey in the garter snake, *Thamnophis elegans*. *Evolution 35*:489–509.

Ashburner, M., Carson, H. L., and Thompson, J. N. Jr. 1981. *The Genetics and Biology of Drosophila*, Vol. 3. New York: Academic Press.

229

Asman, S. M., McDonald, P. T., and Prout, T. 1981. Field studies of genetic control systems for mosquitoes. *Annu. Rev. Genet.* 26:289–318.

Atkinson, W. D. 1979. A comparison of the reproductive strategies of domestic species of *Drosophila. J. Anim. Ecol.* 48:53–64.

– 1981. An ecological interaction between citrus fruit, *Penicillium* moulds and *Drosophila immigrans* Sturtevant (Diptera: Drosophilidae). *Ecol. Entomol.* 6:339–44.

Atkinson, W. D., and Miller, J. A. 1980. Lack of habitat choice in a natural population of *Drosophila subobscura. Heredity* 44:193–9.

Atkinson, W. D., and Shorrocks, B. 1977. Breeding site specificity in the domestic species of *Drosophila. Oecologia (Berlin)* 29:223–32.

Ayala, F. J., and Dobzhansky, Th. 1974. A new subspecies of *Drosophila pseudoobscura. Pan Pacific Entomol.* 50:211–19.

Ayala, F. J., Powell, J. R., and Dobzhansky, Th. 1971. Polymorphisms in continental and island populations of *D. willistoni. Proc. Natl. Acad. Sci. USA* 68:2480–3.

Baker, H. G. 1955. Self-compatibility and establishment after "long-distance" dispersal. *Evolution* 9:347–8.

– 1965. Characteristics and modes of origin of weeds. In *The Genetics of Colonizing Species* (edited by H. G. Baker, and G. L. Stebbins), pp.147–72. New York: Academic Press.

Baker, H. G., and Stebbins, G. L. 1965. *The Genetics of Colonizing Species*,pp.1–588. New York: Academic Press.

Band, H. T. 1963. Genetic structure of populations. II. Viabilities and variances of heterozygotes in constant and fluctuating environments. *Evolution* 17:307–19.

– 1975. A survey of isozyme polymorphism in a *Drosophila melanogaster* natural population. *Genetics* 80:761–71.

Band, H. T., and Ives, P.T. 1968. Genetic structure of populations. IV. Summer environmental variables and lethal and semilethal frequencies in a natural population of *Drosophila melanogaster. Evolution* 22:633–41.

Bantock, C. R., and Noble, K. 1973. Variation with altitude and habitat in *Cepaea hortensis* (Mull). *Zool. J. Linn. Soc.* 53:237–52.

Bantock, C. R., and Price, D. J. 1975. Marginal populations of *Cepaea nemoralis* (L.) on the Brendon Hills, England. I. Ecology and ecogenetics. *Evolution* 29:267–77.

Barclay, H. J., and Gregory, P. T. 1981. An experimental test of models predicting life-history characteristics. *Am. Nat.* 117:944–61.

Barker, J. S. F., and Mulley, J. C. 1976. Isozyme variation in natural populations of *Drosophila buzzatii. Evolution* 30:213–33.

Basden, E. B. 1972. The hymenopterous parasites of the Drosophilidae. *Drosoph. Inf. Serv.* 48:70–2.

Bateman, A. J. 1948. Intra-sexual selection in *Drosophila. Heredity* 2:349–68.

Bateman, M. A. 1967. Adaptations to temperature in geographic races of the Queensland fruit fly, *Dacus (Strumeta) tryoni. Aust. J. Zool.* 15:1141–61.

– 1977. Dispersal and species interaction as factors in the establishment and success of tropical fruit flies in new areas. *Proc. Ecol. Soc. Aust.* 10:106–12.

Bawa, K. S. 1980. Evolution of dioecy in flowering plants. *Annu. Rev. Ecol. Syst. 11*:15–39.

Beardmore, J. A. 1961. Diurnal temperature fluctuation and genetic variance in *Drosophila* populations. *Nature (London) 189*:162–3.

Begg, M., and Hogben, L. 1946. Chemoreceptivity of *Drosophila melanogaster*. *Proc. Roy. Soc. London Ser. B 133*:1–19.

Birch, L. C. 1965. Evolutionary opportunity for insects and mammals in Australia. In *The Genetics of Colonizing Species* (edited by H. G. Baker and G. L. Stebbins), pp. 197–211. New York: Academic Press.

Birch, L. C., Dobzhansky, Th., Elliott, P. O., and Lewontin, R. C. 1963. Relative fitness of geographic races of *Drosophila serrata*. *Evolution 17*:72–83.

Birdsell, J. B. 1973. A basic demographic unit. *Curr. Anthropol. 14*:337–56.

Black, J. N. 1922–52. *Flora of South Australia*. Adelaide: Government Printer.

Bock, I. R., and Parsons, P. A. 1978. The subgenus *Scaptodrosophila* (Diptera: Drosophilidae). *Syst. Entomol. 3*:91–102.

Borowsky, R. 1977. Detection of the effects of selection on protein polymorphisms in natural populations by means of a distance analysis. *Evolution 31*:341–6.

Bowdler, S. 1977. The coastal colonisation of Australia. In *Sunda and Sahul: Prehistoric Studies in Southeastern Asia, Melanesia and Australia* (edited by J. Allen, J. Golson, and R. Jones), pp. 205–46. New York: Academic Press.

Bowler, J. M. 1976. Recent developments in reconstructing late Quaternary environments in Australia. In *The Origin of the Australians* (edited by R. L. Kirk and A. G. Thorne), pp. 55–77. Canberra: Australian Institute of Aboriginal Studies.

Bradbury, J. W. 1981. The evolution of leks. In *Natural Selection and Social Behavior* (edited by R. D. Alexander and D. Tinkle). New York: Chiron Press.

Brittnacher, J. G. 1981. Genetic variation and genetic load due to the male reproductive component of fitness in *Drosophila*. *Genetics 97*:719–30.

Brncic, D. 1970. Studies on the evolutionary biology of Chilean species of *Drosophila*. In *Essays in Evolution and Genetics in Honor of Theodosius Dobzhansky* (edited by M. K. Hecht and W. M. Steere), pp. 401–36. New York: Appleton-Century-Crofts.

Brncic, D., and Budnik, M. 1980. Colonization of *Drosophila subobscura* Colin in Chile. *Drosoph. Inf. Serv. 55*:20.

Brncic, D., and Dobzhansky, Th. 1957. The southernmost Drosophilidae. *Am. Nat. 91*:127–8.

Brough Smyth, R. 1878. *The Aborigines of Victoria with Notes Relating to the Habits of the Natives of Other Parts of Australia and Tasmania*, Vol. 1. Melbourne: Government Printer.

Brown, A. H. D. 1978. Isozymes, plant population genetic structure and genetic conservation. *Theor. Appl. Genet. 52*:145–57.

Brown, A. H. D., and Marshall, D. R. 1981. Evolutionary changes accompanying colonization in plants. In *Evolution Today: Proceedings of the Second International Congress of Systematic and Evolutionary Biology*

232   References

(edited by G. G. E. Scudder and J. L. Reveal), pp. 351–63. Pittsburgh: Carnegie-Mellon University.

Brown, R. G. B. 1965. Courtship behavior in the *Drosophila obscura* group. II. Comparative studies. *Behaviour 25*:281–323.

Bryant, E. H., Van Dijk, H., and van Delden, W. 1981. Genetic variability of the face fly, *Musca autumnalis* de Geer, in relation to a population bottleneck. *Evolution 35*:872–81.

Bush, G. L. 1975. Modes of animal speciation. *Annu. Rev. Ecol. Syst. 6*:339–64.

– 1978. Planning a rational quality control problem for the screwworm fly. In *The Screwworm Problem* (edited by R. H. Richardson), pp. 37–47. Austin: University of Texas Press.

Bushland, R. C. 1975. Screwworm research and eradication. *Bull. Entomol. Soc. Am. 21*:23–6.

Cameron, R. A. D. 1970. The survival, weight-loss and behaviour of three species of land snail in conditions of low humidity. *J. Zool. 160*:143–57.

Capek, M. 1965. The classification of Braconidae (Hym.) in relation to host specificity. *12th Int. Cong. Entomol. London, Abstracts*, pp. 98–9.

Carson, H. L. 1955. The genetic characteristics of marginal populations of *Drosophila. Cold Spring Harbor Symp. Quant. Biol. 20*:276–87.

– 1958. Response to selection under different conditions of recombination in *Drosophila. Cold Spring Harbor Symp. Quant. Biol. 23*:291–306.

– 1965. Chromosomal morphism in geographically widespread species of *Drosophila*. In *The Genetics of Colonizing Species* (edited by H. G. Baker, and G. L. Stebbins), pp. 503–31. New York: Academic Press.

– 1974. Three flies and three islands: parallel evolution in *Drosophila. Proc. Natl. Acad. Sci. USA 71*:3517–21.

– 1977. The unit of genetic change in adaptation and speciation. *Ann. Bot. Gard. 63*:210–23.

– 1978. Speciation and sexual selection in Hawaiian *Drosophila*. In *Ecological Genetics: The Interface* (edited by P. F. Brussard), pp. 93–107. New York: Springer-Verlag.

Carson, H. L., Hardy, D. E., Speith, H. T., and Stone, W. S. 1970. The evolutionary biology of the Hawaiian Drosophilidae. In *Essays in Evolution and Genetics in Honor of Theodosius Dobzhansky* (edited by M. K. Hecht and W. C. Steere), pp. 437–543. New York: Appleton-Century-Crofts.

Carson, H. L., and Ohta, A. T. 1981. Origin of the genetic basis of colonizing ability. In *Evolution Today. Proceedings of the Second International Congress of Systematic and Evolutionary Biology* (edited by G. G. E. Scudder and J. L. Reveal), pp. 365–70. Pittsburg: Carnegie-Mellon University.

Carson, H. L., Val, F. C., Simon, C. M., and Archie, J. W. 1982. Morphometric evidence for incipient speciation in *Drosophila silvestris* from the island of Hawaii. *Evolution 36*:132–40.

Carton, Y. 1977. Attraction de *Cothonaspis* sp. (Hymenoptère Cynipidae) par le milieu trophique de son hote: *Drosophila melanogaster*. In *Comportement des insectes et milieu trophique. Colloq. Int. CNRS 265*:285–303.

Carton, Y., and Kitano, H. 1981. Evolutionary relationships to parasitism by seven species of the *Drosophila melanogaster* subgroup. *Biol. J. Linn. Soc. 16*:227–41.

Caspari, E. W. 1952. Pleiotropic gene action. *Evolution 6*:1–18.

– 1977. Genetic mechanisms and behavior. In *Genetics, Environment and Intelligence* (edited by A. Oliverio), pp. 3–22. Amsterdam: North Holland.

Cavalli, L. L., and Maccacaro, G. A. 1952. Polygenic inheritance of drug-resistance in the bacterium *Escherichia coli*. *Heredity 6*:311–31.

Cavalli-Sforza, L. L., and Feldman, M. W. 1981. *Cultural Transmission and Evolution: A Quantitative Approach*. Princeton; N.J.: Princeton University Press.

Chabora, P. C., Smolin, S. J., and Kopelman, A. K. 1979. The life history of *Pseudeucoila* sp., a protelian parasite of *Drosophila*. *Ann. Entomol. Soc. Am. 72*:495–9.

Chew, F. S. 1980. Foodplant preferences of *Pieris* caterpillars (Lepidoptera). *Oecologia 46*:347–53.

Christensen, B. 1977. Habitat preference among amylase genotypes in *Asellus aquaticus* (Isopoda, Crustacea). *Hereditas 87*:21–6.

Clarke, B. 1975. A contribution of ecological genetics to evolutionary theory: detecting the direct effects of natural selection on particular polymorphic loci. *Genetics 79*:101–13.

– 1979. The evolution of genetic diversity. *Proc. Roy. Soc. London Ser. B 205*:453–74.

Clarke, J. M., Maynard Smith, J., and Sondhi, K. C. 1961. Asymmetrical response to selection for rate of development in *Drosophila subobscura*. *Genet. Res. 2*:70–81.

Clegg, M. T., Allard, R. W., and Kahler, A. L. 1972. Is the gene the unit of selection? Evidence from two experimental plant populations. *Proc. Natl. Acad. Sci. USA 69*:2474–8.

Coen, E. S., Thoday, J. M., and Dover, G. 1982. Rate of turnover of structural variants in the rDNA gene family of *Drosophila melanogaster*. *Nature (London) 295*:564–8.

Cohet, Y., and David, J. R. 1980. Geographic divergence and sexual behaviour: comparison of mating systems in French and Afrotropical populations of *Drosophila melanogaster*. *Genetica 54*:161–5.

Cohet, Y., Vouidibio, J., and David, J. R. 1980. Thermal tolerance and geographic distribution: a comparison of cosmopolitan and tropical endemic *Drosophila* species. *J. Therm. Biol. 5*:69–74.

Cole, L. C. 1954. The population consequences of life history phenomena. *Quart. Rev. Biol. 29*:103–37.

Coluzzi, M., Sabatini, A., Petrarca, V., and DiDeco, M. A. 1979. Chromosomal differentiation and adaptation to human environments in the *Anopheles gambiae* complex. *Trans. Roy. Soc. Trop. Med. Hyg. 73*:483–97.

Conway, G. 1981. Man versus pests. In *Theoretical Ecology: Principles and Applications* (edited by R. M. May), pp. 356–86. Sunderland, Mass.: Sinauer Associates.

Cook, R. M., Parsons, P. A., and Bock, I. R. 1977. Australian endemic *Dro-*

sophila. II. A new *Hibiscus*-breeding species with its description. *Aust. J. Zool.* 25:755–63.

Crovello, T. J., and Hacker, C. S. 1972. Evolutionary strategies in life table characteristics among feral and urban strains of *Aedes aegypti* (L.). *Evolution* 26:185–96.

Crow, J. F. 1957. Genetics of insect resistance to chemicals. *Annu. Rev. Entomol.* 2:227–46.

Crumpacker, D. W., and Williams, J. S. 1973. Density, dispersion and population structure in *Drosophila pseudoobscura*. *Ecol. Monogr.* 43:499–538.

Curr, E. M. 1886. *The Australian Race*, Vol. 1. Melbourne: Government Printer.

da Cunha, A. B., Burla, H., and Dobzhansky, Th. 1950. Adaptive chromosomal polymorphism in *Drosophila willistoni*. *Evolution* 4:212–35.

da Cunha, A. B., and Dobzhansky, Th. 1954. A further study of chromosomal polymorphism in *Drosophila willistoni* in its relation to the environment. *Evolution* 8:119–34.

Darlington, P. J. 1957. *Zoogeography: The Geographical Distribution of Animals*. New York: Wiley.

Darwin, C. 1859. *On the Origin of Species by Means of Natural Selection*. London: Murray.

David, J., and Bocquet, C. 1975. Similarities and differences in latitudinal adaptation of two *Drosophila* sibling species. *Nature (London)* 257:588–90.

David, J., Bocquet, C., and Pla, E. 1976. New results on the genetic characteristics of the Far East race of *Drosophila melanogaster*. *Genet. Res.* 28:253–60.

David, J. R., van Herrewege, J., Monclus, M., and Prevosti, A. 1979. High ethanol tolerance in two distantly related *Drosophila* species: a probable case of recent convergent adaptation. *Comp. Biochem. Physiol. C* 63:53–6.

Day, T. H., Hillier, P. C., and Clarke, B. 1974. Properties of genetically polymorphic isozymes of alcohol dehydrogenase in *Drosophila melanogaster*. *Biochem. Genet.* 11:141–52.

Decker, G. C., and Bruce, W. N. 1952. House fly resistance to chemicals. *Am. J. Trop. Med. Hyg.* 1:395–403.

Deltombe-Lietaert, M. C., Delcour, J., Lenelle-Montfort, N., and Elens, A. 1979. Ethanol metabolism in *Drosophila melanogaster*. *Experientia* 35:579–81.

Denno, R. F., and Dingle, H. 1981. *Insect Life History Patterns: Habitat and Geographic Variation*. New York: Springer-Verlag.

Derr, J. A. 1980. The nature of variation in life history characters of *Dysdercus bimaculatus* (Heteroptera: Pyrrhocoridae), a colonizing species. *Evolution* 34:548–57.

Derr, J. A., Alden, B., and Dingle, H. 1981. Insect life histories in relation to migration, body size, and host plant array: a comparative study of *Dysdercus*. *J. Anim. Ecol.* 50: 181–93.

Desforges, M. F., and Wood-Gush, D. G. M. 1975a. A behavioural comparison

of domestic and mallard ducks. Habituation and flight reactions. *Anim. Behav. 23*:692–7.
- 1975b. A behavioural comparison of domestic and mallard ducks. Spatial relationships in small flocks. *Anim. Behav. 23*:698–705.
- 1976. Behavioural comparison of Aylesbury and Mallard ducks: sexual behaviour. *Anim. Behav. 24*:391–7.
de Souza, H. M. L., da Cunha, A. B., and dos Santos, E. P. 1970. Adaptive polymorphism of behavior evolved in laboratory populations of *Drosophila willistoni*. *Am. Nat. 104*:175–89.
Dethier, V. G. 1954. Evolution of feeding preferences in phytophagous insects. *Evolution 8*:33–54.
Dethier, V. G., Barton-Browne, L., and Smith, C. N. 1960. The designation of chemicals in terms of the responses they elicit from insects. *J. Econ. Entomol. 53*:134–6.
Diamond, J. M. 1973. Distributional ecology of New Guinea birds. *Science 179*:759–69.
- 1975. Assembly of species communities. In *Ecology and Evolution of Communities* (edited by M. L. Cody and J. M. Diamond), pp. 342–444. Cambridge; Mass.: Harvard University Press.
Diamond, J. M., and Marshall, A. G. 1977. Niche shifts in New Hebridean birds. *Emu 77*:61–72.
Dickinson, W. J. 1980. Evolution of patterns of gene expression in Hawaiian picture winged *Drosophila*. *J. Mol. Evol. 16*:73–94.
Dingle, H. 1974. The experimental analysis of migration and life history strategies in insects. In *Experimental Analysis of Insect Behaviour* (edited by L. Barton Browne), pp. 329–42. New York: Springer-Verlag.
- 1981. Geographic variation and behavioral flexibility in milkweed bug life histories. In *Insect Life History Patterns: Habitat and Geographic Variation* (edited by R. F. Denno and H. Dingle), pp. 57–73. New York: Springer-Verlag.
Dobzhansky, Th. 1950. Evolution in the tropics. *Am. Sci. 38*:209–21.
- 1965. 'Wild' and 'domestic' species of *Drosophila*. In *The Genetics of Colonizing Species* (edited by H. G. Baker and G. L. Stebbins), pp. 533–46. New York: Academic Press.
Dobzhansky, Th., Hunter, A. S., Pavlovsky, O., Spassky, B., and Wallace, B. 1963. Genetics of natural populations. XXXI. Genetics of an isolated marginal population of *Drosophila pseudoobscura*. *Genetics 48*:91–103.
Dobzhansky, Th., and Powell, J. R. 1975a. The *willistoni* group of sibling species of *Drosophila*. In *Handbook of Genetics* (edited by R. C. King), Vol. 3, pp. 589–622. New York: Plenum Press.
- 1975b. *Drosophila pseudoobscura* and its American relatives, *Drosophila persimilis* and *Drosophila miranda*. In *Handbook of Genetics* (edited by R. C. King), Vol. 3, pp. 537–87. New York: Plenum Press.
Dobzhansky, Th., and Spassky, B. 1969. Artificial and natural selection for two behavioral traits in *Drosophila pseudoobscura*. *Proc. Natl. Acad. Sci. USA 62*:75–80.
Dobzhansky, Th., and Wright, S. 1943. Genetics of natural populations. X. Dispersion rates in *Drosophila pseudoobscura*. *Genetics 28*:304–40.

Douglas, M. M., and Grula, J. W. 1978. Thermoregulatory adaptations allow-
ing ecological range expansion by the Pierid butterfly, *Nathalis iole* Bios-
duval. *Evolution 32*:776–83.

Doyle, R. W. 1976. Analysis of habitat loyalty and habitat preferences in the
settlement behavior of planktonic marine larvae. *Am. Nat. 110*:719–30.

Dubinin, N. P., and Tiniakov, G. G. 1947. Inversion gradients and selection
in ecological races of *Drosophila funebris*. *Am. Nat. 81*:148–53.

Dudai, Y. 1977. Properties of learning and memory in *Drosophila melanogaster*. *J. Comp. Physiol. 114*:69–89.

Edmunds, G. F., and Alstad, D. N. 1981. Responses of black pineleaf scales
to host plant variability. In *Insect Life History Patterns: Habitat and Geographic Variation* (edited by R. F. Denno and H. Dingle), pp. 29–38. New
York: Springer-Verlag.

Edwards, A. W. F. 1962. Migrational selection. *Heredity 18*:101–6.

Ehrlich, P. R., Murphy, D. D., Singer, M. C., Sherwood, C. B., White, R.
R., and Brown, I. L. 1980. Extinction, reduction, stability and increase:
the responses of checkerspot butterfly (*Euphydryas*) populations to the
California drought. *Oecologia 46*:101–5.

Ehrlich, P. R., and Raven, P. H. 1969. Differentiation of populations. *Science
165*:1228–32.

Ehrman, L. 1964. Genetic divergence in M. Vetukhiv's experimental populations of *Drosophila pseudoobscura* I. Rudiments of sexual isolation.
Genet. Res. 5:150–7.

Ehrman, L., and Parsons, P. A. 1981a. *Behavior Genetics and Evolution*, 450
pp. New York: McGraw-Hill.

– 1981b. Sexual isolation among isofemale strains within a population of *Drosophila immigrans*. *Behav. Genet. 11*:127–33.

Ehrman, L., and Probber, J. 1978. Rare *Drosophila* males: the mysterious
matter of choice. *Am. Sci. 66*:216–22.

Eldredge, N., and Gould, S. J. 1972. Punctuated equilibria: an alternative to
phyletic gradualism. In *Models in Paleobiology* (edited by T. J. M.
Schopf), pp. 82–115. San Francisco: Freeman & Cooper.

Elton, C. S. 1958. *The Ecology of Invasions by Animals and Plants*, pp. 1–
181. London: Methuen.

Endler, J. A. 1979. Gene flow and life history patterns. *Genetics 93*:263–
84.

Falconer, D. S. 1960. *Introduction to Quantitative Genetics*. London: Long-
man.

Feeny, P. 1975. Biochemical coevolution between plants and their insect herbivores. In *Coevolution of Animals and Plants* (edited by L. E. Gilbert
and P. H. Raven), pp. 3–19. Austin: University of Texas Press.

Fenner, F. 1965. Myxoma virus and *Oryctolagus cuniculus*: two colonizing
species. In *The Genetics of Colonizing Species* (edited by H. G. Baker
and G. L. Stebbins), pp. 485–99. New York: Academic Press.

– 1977. Processes that contribute to the establishment and success of exotic
microorganisms in Australia. *Proc. Ecol. Soc. Aust. 10*:39–61.

Fincham, J. R. S. 1972. Heterozygous advantage as a likely general basis for
enzyme polymorphisms. *Heredity 28*:387–91.

Finnerty, V., and Johnson, G. 1979. Post-translational modification as a potential explanation of high levels of enzyme polymorphism: xanthine dehydrogenase and aldehyde oxidase in *Drosophila melanogaster*. *Genetics* 91:695–722.

Fisher, R. A. 1918. The correlation between relative on the supposition of Mendelian inheritance. *Trans. Roy. Soc. Edinburgh* 52:399–433.

— 1930. *The Genetical Theory of Natural Selection*. Oxford University Press (Clarendon Press).

— 1936. The use of multiple measurements in taxonomic problems. *Ann. Eugenics* 7:179–88.

— 1937. The wave of advance of advantageous genes. *Ann. Eugenics* 7:355–69.

Fletcher, B. S. 1974. The ecology of a natural population of the Queensland fruit fly, *Dacus tryoni*. V. The dispersal of adults. *Aust. J. Zool.* 22:189–202.

Fontdevila, A., Ruiz, A., Alonso, G., and Ocaña, J. 1981. Evolutionary history of *Drosophila buzzatii*. I. Natural chromosomal polymorphism in colonized populations of the old world. *Evolution* 35:148–57.

Foster, G. G., Whitten, M. J., Prout, T., and Gill, R. 1972. Chromosome rearrangements for the control of insect pests. *Science* 176:875–80.

Frankel, O. H., and Soulé, M. E. 1981. *Conservation and Evolution*. Cambridge University Press.

Fuller, J. L., and Thompson, W. R. 1960. *Behavior Genetics*. New York: Wiley.

Futuyma, D. J., and Mayer, G. C. 1980. Non-allopatric speciation in animals. *Syst. Zool.* 29:254–71.

Fuyama, Y. 1976. Behavior genetics of olfactory responses in *Drosophila*. I. Olfactometry and strain differences in *Drosophila melanogaster*. *Behav. Genet.* 6:407–20.

— 1978. Behavior genetics of olfactory responses in *Drosophila*. II. An odorant-specific variant in a natural population of *Drosophila melanogaster*. *Behav. Genet.* 8:399–414.

Garten, C. J. 1976. Relationships between aggressive behavior and genic heterozygosity in the Oldfield mouse, *Peromyscus polionotus*. *Evolution* 30:59–72.

Gibson, J. B. 1970. Enzyme flexibility in *Drosophila melanogaster*. *Nature* (*London*) 227:959–60.

Gibson, J. B., and Oakeshott, J. G. 1980. *Genetic Studies of Drosophila Populations*, pp. 1–27. Canberra: Australian National University Press.

Giesel, J. T. 1974. *The Biology and Adaptability of Natural Populations*, pp. 1–177. St. Louis: Mosby.

— 1979. Genetic co-variation of survivorship and other fitness indices in *Drosophila melanogaster*. *Exp. Gerontol.* 14:323–8.

Giesel, J. T., and Zettler, E. E. 1980. Genetic correlations of life historical parameters and certain fitness indices in *Drosophila melanogaster*: r_m, r_s, diet breadth. *Oecologia* 47:299–302.

Gould, F. 1979. Rapid host range evolution in a population of the phytophagous mite *Tetranychus urticae* Koch. *Evolution* 33:791–802.

Gould, S. J., and Johnston, R. F. 1972. Geographic variation. *Annu. Rev. Ecol. Syst. 3*:457–98.

Gould, S. J., and Lewontin, R. C. 1979. The spandrels of San Marco and the Panglossian paradigm: a critique of the adaptationist programme. *Proc. Roy. Soc. London Ser. B 205*:581–98.

Graham, J. B., Rubinoff, I., and Hecht, M. K. 1971. Temperature physiology of the sea snake *Pelamis platurus*: an index of its colonization potential in the Atlantic Ocean. *Proc. Natl. Acad. Sci. USA 68*:1360–3.

Grant, M. C., and Antonovics, J. 1978. Biology of ecologically marginal populations of *Anthoxanthum odoratum* L. Phenetics and dynamics. *Evolution 32*:822–38.

Grant, V. 1963. *The Origin of Adaptations*. New York: Columbia University Press.

– 1975. *Genetics of Flowering Plants*. New York: Columbia University Press.

Greaves, J. H., and Rennison, B. D. 1973. Population aspects of warfarin resistance in the Brown rat, *Rattus norvegicus*. *Mammal. Rev. 3*:27–9.

Griffing, J. B. 1956. Concept of general and specific combining ability in relation to diallel crossing systems. *Aust. J. Biol. Sci. 9*:463–93.

Grossfield, J. 1966. The influence of light on the mating behavior of *Drosophila*. *Univ. Tex. Publ. 6615*:147–76.

– 1971. Geographic distribution and light-dependent behavior in *Drosophila*. *Proc. Natl. Acad. Sci. USA 68*:2669–73.

– 1972. The use of behavioral mutants in biological control. *Behav. Genet. 2*:311–19.

Grossfield, J., and Parsons, P. A. 1975. Locality records for some endemic Australian Drosophilidae based on winter collections. *Proc. Roy. Soc. Vict. 87*:235–8.

Gruneberg, H. 1963. *The Pathology of Development*. New York: Wiley.

Guhl, A. M. 1962. The behaviour of chickens. In *The Behaviour of Domestic Animals* (edited by E. S. E. Hafez), pp. 491–530. Paris: Baillière.

Guhl, A. M., Craig, J. V., and Mueller, C. D. 1960. Selective breeding for aggressiveness in chickens. *Poult. Sci. 39*:970–80.

Hadler, N. M. 1964. Heritability and phototaxis in *Drosophila melanogaster*. *Genetics 50*:1269–77.

Hale, E. B. 1962. Domestication and the evolution of behaviour. In *The Behavior of Domestic Animals* (edited by E. S. E. Hafez), 2nd ed., pp. 22–42. Baltimore: Williams & Wilkins.

Harper, J. L. 1977. *Population Biology of Plants*, 892 pp. New York: Academic Press.

Harrison, R. A. 1959. Acalypterate Diptera of New Zealand. *N.Z. Dept. Sci. Ind. Res. Bull. 128.*

Heed, W. B., and Heed, S. R. 1972. Ecology, weather and dispersal of *Drosophila* on an island mountain. *Drosoph. Inf. Serv. 48*:100–1.

Heron, A. C. 1972a. Population ecology of a colonizing species: the pelagic tunicate *Thalia democratica*. I. Individual growth rate and generation time. *Oecologia 10*:269–93.

– 1972b. Population ecology of a colonizing species: The pelagic tunicate *Thalia democratica*. II. Population growth rate. *Oecologia 10*:294–312.

Hershberger, W. A., and Smith, M. P. 1967. Conditioning in *Drosophila melanogaster*. *Anim. Behav. 15*:259–62.

Hirsch, J., and Boudreau, J. 1958. Studies in experimental behavior genetics. The heritability of phototaxis in a population of *Drosophila melanogaster*. *J. Comp. Physiol. Psychol. 51*:647–51.

Hoenigsberg, H. F., Castro, L. E., Granobles, L. A., and Idrobo, J. M. 1969. Population genetics in the American tropics. II. The comparative genetics of *Drosophila* in European and neotropical environments. *Genetica 40*:43–60.

Hoffman, R. J. 1978. Environmental uncertainty and evolution of physiological adaptation in *Colias* butterflies. *Am. Nat. 112*:999–1015.

Holdren, C. E., and Ehrlich, P. R. 1981. Long range dispersal in checkerspot butterflies: transplant experiments with *Euphydryas gillettii. Oecologia 50*:125–9.

Holmes, R. S., Moxon, L. N., and Parsons, P. A. 1980. Genetic variability of alcohol dehydrogenase among Australian *Drosophila* species: correlation of ADH biochemical phenotype with ethanol resource utilization. *J. Exp. Zool. 214*:199–204.

Hosgood, S. M. W., MacBean, I. T., and Parsons, P. A. 1968. Genetic heterogeneity and accelerated responses to directional selection in *Drosophila. Mol. Gen. Genet. 101*:217–26.

Hosgood, S. M. W., and Parsons, P. A. 1966. Differences between *D. simulans* and *D. melanogaster* in tolerances to laboratory temperatures. *Drosoph. Inf. Serv. 41*:176.

– 1967. The exploitation of genetic heterogeneity among the founders of laboratory populations of *Drosophila* prior to directional selection. *Experientia 23*:1066–7.

– 1968. Polymorphism in natural populations of *Drosophila* for the ability to withstand temperature shocks. *Experientia 24*:727–8.

– 1971. Genetic heterogeneity among the founders of laboratory populations of *Drosophila*. IV. Scutellar chaetae in different environments. *Genetica 42*:42–52.

Hoy, M. A. 1978. Variability in diapause attributes of insects and mites: some evolutionary and practical implications. In *Proceedings in Life Sciences. Evolution of Insect Migration and Diapause* (edited by H. Dingle), pp. 101–26. New York: Springer-Verlag.

Hutner, S. H., Kaplan, H. M., and Enzmann, E. V. 1937. Chemicals attracting *Drosophila. Am. Nat. 71*:575–81.

Istock, C. A. 1981. Natural selection and life history variation: theory plus lessons from a mosquito. In *Insect Life History Patterns: Habitat and Geographic Variation* (edited by R. F. Denno, and H. Dingle), pp. 113–27. New York: Springer-Verlag.

Istock, C. A., Vavra, K. J., and Zimmer, H. 1976. Ecology and evolution of the pitcher-plant mosquito. 3. Resource tracking by a natural population. *Evolution 30*:548–57.

Istock, C. A., Wasserman, S. S., and Zimmer, H. 1975. Ecology and evolution in the pitcher-plant mosquito. 1. Population dynamics and laboratory responses to food and population density. *Evolution 29*:296–312.

Istock, C. A., Zisfein, J., and Vavra, K. J. 1976. Ecology and evolution of the pitcher-plant mosquito. 2. The substructure of fitness. *Evolution 30*:535–47.

Jacobs, M. E. 1960. The influence of light on mating of *D. melanogaster*. *Ecology 41*:182–8.

Jaenike, J. 1981. Criteria for ascertaining the existence of host races. *Am.Nat. 117*:830–4.

Jaenike, J., and Selander, R. K. 1979. Ecological generalism in *Drosophila falleni*: genetic evidence. *Evolution 33*:741–8.

Jain, S. K. 1976. The evolution of inbreeding in plants. *Annu. Rev. Ecol. Syst. 7*:469–95.

– 1979. Adaptive strategies: polymorphism, plasticity, and homeostasis. In *Topics in Plant Population Biology* (edited by O. T. Solbrig, S. Jain, G. B. Johnson, and P. H. Raven), pp. 160–87. New York: Columbia University Press.

Johnson, G. B. 1974. Enzyme polymorphism and metabolism. *Science 184*:28–37.

– 1979a. Enzyme polymorphism: genetic variation in the physiological phenotype. In *Topics in Plant Population Biology* (edited by O. T. Solbrig, S. Jain, G. B. Johnson, and P. H. Raven), pp. 62–83. New York: Columbia University Press.

– 1979b. Genetic polymorphism among enzyme loci. In *Physiological Genetics* (edited by J. G. Scandalios), pp. 239–73. New York: Academic Press.

Johnston, J. S., and Heed, W. B. 1976. Dispersal of desert-adapted *Drosophila*: the saguaro-breeding *D. nigrospiracula*. *Am. Nat. 110*:629–51.

Johnston, R. F., and Klitz, W. J. 1977. Variation and evolution in a granivorous bird: the house sparrow. In *Granivorous Birds in Ecosystems* (edited by J. Pinowski and S. C. Kendeigh), pp. 15–51. Cambridge University Press.

Jones, J. S. 1973a. Ecological genetics of a population of the snail *Cepaea nemoralis* at the northern limit of its range. *Heredity 31*:201–11.

– 1973b. The genetic structure of a southern peripheral population of the snail *Cepaea nemoralis*. *Proc. Roy. Soc. London Ser. B 183*:371–84.

– 1981. An uncensored page of fossil history. *Nature (London) 293*:427–8.

Jones, J. S., Bryant, S. H., Lewontin, R. C., Moore, J. A., and Prout, T. 1981. Gene flow and the geographical distribution of a molecular polymorphism in *Drosophila pseudoobscura*. *Genetics 98*:157–78.

Jones, J. S., Leith, B. H., and Rawlings, P. 1977. Polymorphism in *Cepaea*: a problem with too many solutions? *Annu. Rev. Ecol. Syst. 8*:109–43.

Jones, J. S., and Probert, R. F. 1980. Habitat selection maintains a deleterious allele in a heterogeneous environment. *Nature (London) 287*:632–3.

Jones, J. S., Selander, R. K., and Schnell, G. D. 1980. Patterns of morphological and molecular polymorphism in the land snail *Cepaea nemoralis*. *Biol. J. Linn. Soc. 14*:359–87.

Jones, R. 1977. Man as an element of a continental fauna: The case of the sundering of the Bassian bridge. In *Sunda and Sahul: Prehistoric Studies in Southeast Asia, Melanesia and Australia* (edited by J. Allen, J. Golson, and R. Jones), pp. 317–86. New York: Academic Press.

Kambysellis, M. P., and Heed, W. B. 1971. Studies of oogenesis in natural populations of Drosophilidae. I. Relation of ovarian development and ecological habitats of the Hawaiian species. *Am. Nat. 105*:31–49.

Kamping, A., and van Delden, W. 1978. The alcohol dehydrogenase polymorphism in populations of *Drosophila melanogaster*. 2. The relation between ADH activity and adult mortality. *Biochem. Genet. 16*:541–51.

Kaul, D., and Parsons, P. A. 1966. Competition between males in the determination of mating speed in *Drosophila pseudoobscura*. *Aust. J. Biol. Sci. 19*:945–7.

Kawanishi, M., and Watanabe, T. K. 1978. Differences in photopreference as a cause of coexistence of *Drosophila simulans* and *D. melanogaster* in nature. *Jap. J. Genet. 53*:209–14.

Keeler, K. H. 1978. Intra-population differentiation in annual plants. II. Electrophoretic variation in *Veronica peregrina*. *Evolution 32*:638–45.

Kekić, V., Taylor, C. E., and Andjelković, M. 1980. Habitat choice and resource specialization by *Drosophila subobscura*. *Genetika 12*:219–25.

Kellogg, D. E. 1975. The role of phyletic change in the evolution of *Pseudocubus vema* (Radiolaria). *Paleobiology 1*:359–70.

Kempthorne, O. 1957. *An Introduction to Genetic Statistics*. New York: Wiley.
– 1981. Book review of *Quantitative Genetic Variation* (edited by J. N. Thompson, Jr., and J. M. Thoday), New York: Academic Press. *Social Biol. 27*:241–6.

Kettlewell, H. B. D. 1973. *The Evolution of Melanism*. New York: Oxford University Press.

Kilias, G., Alahiotis, S. N., and Pelecanos, M. 1980. A multifactorial genetic investigation of speciation theory using *Drosophila melanogaster*. *Evolution 34*:730–7.

King, J. A. 1967. Behavioral modification of the gene pool. In *Behavior-Genetic Analysis* (edited by J. Hirsch), pp. 22–43. New York: McGraw-Hill.

King, J. C., and Sømme, L. 1958. Chromosomal analysis of the genetic factors for resistance to DDT in two resistant lines of *Drosophila melanogaster*. *Genetics 43*:577–93.

Kirk, R. L., and Thorne, A. G. (eds.) 1976. *The Origin of the Australians*. Canberra: Australian Institute of Aboriginal Studies.

Kloet, G. S., and Hincks, W. D. 1945. *A Check List of British Insects*. Stockport: Kloet & Hincks.

Lachaise, D. 1974. Les Drosophilides des savanes préforestières de la région tropical de Lamto (Côte-d'Ivoire). V. Les régimes alimentaires. *Ann. Soc. Entomol. Fr. 10*:3–50.

Lakovaara, S., Saura, A., Lokki, J., and Lankinen, P. 1976. Genic polymorphism in marginal populations of *Drosophila*. *Not. Entomol. 56*:65–72.

Lande, R. 1980. The genetic covariance between characters maintained by pleiotropic mutations. *Genetics 94*:203–15.

Lawton, J. H., and Schroder, D. 1977. Effects of plant type, size of geographical range and taxonomic isolation on number of insect species associated with British plants. *Nature (London) 265*:137–40.

Lee, B. T. O., and Parsons, P. A. 1968. Selection, prediction and response. *Biol. Rev. 43*:139–74.

Lerner, I. M. 1968. *Heredity, Evolution and Society*. San Francisco: Freeman.

Levene, H. 1953. Genetic equilibrium when more than one ecological niche is available. *Am. Nat. 87*:331–3.

Levins, R. 1968. *Evolution in Changing Environments: Some Theoretical Explorations*. Princeton; N.J.: Princeton University Press.

– 1969. Thermal acclimation and heat resistance in *Drosophila* species. *Am. Nat. 103*:483–99.

Levins, R., and MacArthur, R. H. 1966. Maintenance of genetic polymorphism in a heterogeneous environment: variations on a theme by Howard Levene. *Am. Nat. 100*:585–90.

Lewis, D. 1942. The evolution of sex in flowering plants. *Biol. Rev. 17*:46–67.

Lewis, H. 1962. Catastrophic selection as a factor in speciation. *Evolution 16*:257–71.

Lewontin, R. C. 1959. On the anomalous response of *Drosophila pseudoobscura* to light. *Am. Nat. 93*:321–8.

– 1965. Selection for colonizing ability. In *The Genetics of Colonizing Species* (edited by H. G. Baker and G. L. Stebbins), pp. 77–94. New York: Academic Press.

– 1974. *The Genetic Basis of Evolutionary Change*, pp. 1–346. New York: Columbia University Press.

Lewontin, R. C., and Birch, L. C. 1966. Hybridization as a source of variation for adaptation to new environments. *Evolution 20*:315–36.

Lindauer, M. 1975. Evolutionary aspects of orientation and learning. In *Function and Evolution in Behaviour* (edited by G. Baerends, C. Beer, and A. Manning), pp. 228–42. Oxford University Press (Clarendon Press).

Linhart, Y. B. 1974. Intra-population differentiation in annual plants. I. *Veronica peregrina* L. raised under non-competitive conditions. *Evolution 28*:232–43.

Lister, B. C. 1976. The nature of niche expansion in West Indian *Anolis* lizards. II. Evolutionary components. *Evolution 30*:677–92.

Lumme, J. 1978. Phenology and photoperiodic diapause in northern populations of *Drosophila*. In *Proceedings in Life Sciences. Evolution of Insect Migration and Diapause* (edited by H. Dingle), pp. 145–70. New York: Springer-Verlag.

Lynch, C. B., and Hegmann, J. P. 1972. Genetic differences influencing behavioral temperature regulation in small mammals. Nesting by *Mus musculus. Behav. Genet. 1*:43–53.

MacArthur, J. W. 1975. Environmental fluctuations and species diversity. In *Ecology and Evolution of Communities* (edited by M. L. Cody, and J. M. Diamond), pp. 74–80. Cambridge, Mass.: Harvard University Press (Belknap Press).

MacArthur, R. H. 1972. *Geographical Ecology*. New York: Harper & Row.

MacArthur, R. H., and Wilson, E. O. 1967. *The Theory of Island Biogeography*, pp. 1–203. Princeton, N.J.: Princeton University Press.

MacBean, I. T., McKenzie, J. A., and Parsons, P. A. 1971. A pair of closely linked genes controlling high scutellar chaeta number in *Drosophila. Theor. Appl. Genet. 41*:227–35.

Malogolowkin-Cohen, C. H., Simmons, A. S., and Levene, H. 1965. A study of sexual selection between certain strains of *Drosophila paulistorum*. *Evolution 19*:95–103.

Malpica, J. M., and Vassallo, J. M. 1980. A test for the selective origin of environmentally correlated allozyme patterns. *Nature (London)* 286:407–8.

Manning, A. 1967. "Pre-imaginal conditioning" in *Drosophila*. *Nature (London)* 216:338–40.

Marinković, D., Ayala, F. J., and Andjelković, M. 1978. Genetic polymorphism and phylogeny of *Drosophila subobscura*. *Evolution 32*:164–73.

Marinković, D., Tucić, N., and Kekić, V. 1980. Genetic variation and ecological adaptations. *Genetica 52–53*:249–62.

Markow, T. A. 1975. A genetic analysis of phototactic behavior in *Drosophila melanogaster* I. Selection in presence of inversions. *Genetics 79*:527–34.

Marshall, I. D., and Douglas, G. W. 1961. Studies in the epidemiology of infectious myxomatosis of rabbits. VIII. Further observations on changes in the innate resistance of Australian wild rabbits exposed to myxomatosis. *J. Hyg. 59*:117–22.

Mather, K. 1943. Polygenic inheritance and natural selection. *Biol. Rev. 18*:32–64.

– 1955. Polymorphism as an outcome of disruptive selection. *Evolution 9*:52–61.

Mather, K., and Jinks, J. L. 1971. *Biometrical Genetics*. London: Chapman & Hall.

Maynard Smith, J. 1966. Sympatric speciation. *Am. Nat. 100*:637–50.

Maynard Smith, J., and Haigh, J. 1974. The hitch-hiking effect of a favourable gene. *Genet. Res. 23*:23–35.

Mayr, E. 1963. *Animal Species and Evolution*, pp. 1–797. Cambridge, Mass.: Harvard University Press (Belknap Press).

– 1965. Summary. In *The Genetics of Colonizing Species* (edited by H. G. Baker, and G. L. Stebbins), pp. 553–62. New York: Academic Press.

– 1974. Behavior programs and evolutionary strategies. *Am. Sci. 62*:650–9.

– 1977. The study of evolution, historically viewed. In *The Changing Scenes in Natural Sciences 1776-1976*, Special Publication 12, pp.39–58. Philadelphia: Academy of Natural Sciences.

McBride, G. 1958. The environment and animal breeding problems. *Anim. Breed. Abstr. 26*:349–58.

McComb, J. A. 1966. The sex form of species in the flora of the southwest of Western Australia. *Aust. J. Bot. 14*:303–16.

McGuire, T. R., and Hirsch, J. 1977. Behavior-genetic analysis of *Phormia regina*: conditioning, reliable individual differences, and selection. *Proc. Natl. Acad. Sci. USA 74*:5193–7.

McKay, T. F. C., and Doyle, R. W. 1978. An ecological genetic analysis of the settling behaviour of a marine polychaete. I. Probability of settlement and gregarious behaviour. *Heredity 40*:1–12.

McKenzie, J. A. 1974. The distribution of vineyard populations of *Drosophila melanogaster* and *Drosophila simulans* during vintage and non-vintage periods. *Oecologia 14*:1–16.

- 1975. The influence of low temperature on survival and reproduction in populations of *Drosophila melanogaster*. *Aust. J. Zool.* 23:237–47.
- 1976. The release of a compound-chromosome stock in a vineyard population of *Drosophila melanogaster*. *Genetics* 82:685–95.
- 1978. The effect of developmental temperature on population flexibility in *Drosophila melanogaster* and *D. simulans*. *Aust. J. Zool.* 26:105–12.
McKenzie, J. A., and McKechnie, S. W. 1978. Ethanol tolerance and the *Adh* polymorphism in a natural population of *Drosophila melanogaster*. *Nature (London)* 272:75–6.
- 1979. A comparative study of resource utilization in natural populations of *Drosophila melanogaster* and *D. simulans*. *Oecologia* 40:299–309.
McKenzie, J. A., and Parsons, P. A. 1971. Variations in mating propensities in strains of *Drosophila melanogaster* with different scutellar chaeta numbers. *Heredity* 26:313–22.
- 1972. Alcohol tolerance: an ecological parameter in the relative success of *Drosophila melanogaster* and *Drosophila simulans*. *Oecologia* 10:373–88.
- 1974a. Numerical changes ånd environmental utilization in natural populations of *Drosophila*. *Aust. J. Zool.* 22:175–87.
- 1974b. The genetic architecture of resistance to desiccation in populations of *Drosophila melanogaster* and *D. simulans*. *Aust. J. Biol. Sci.* 27:441–56.
- 1974c. Microdifferentiation in a natural population of *Drosophila melanogaster* to alcohol in the environment. *Genetics* 77:385–94.
Meats, A. 1981. The bioclimatic potential of the Queensland fruit fly, *Dacus tryoni*, in Australia. *Proc. Ecol. Soc. Aust.* 11:151–61.
Médioni, J. 1962. Contribution à l'étude psycho-physiologique et génétique du phototropisme d'un insecte: *Drosophila melanogaster* Meigen. Thèse Fac. Sci. Strasbourg.
Meehan, B. 1977. Man does not live by calories alone: The role of shellfish in a coastal cuisine. In *Sunda and Sahul: Prehistoric Studies in Southeast Asia, Melanesia and Australia* (edited by J. Allen, J. Golson, and R. Jones), pp. 493–531. New York: Academic Press.
Merrell, D. J. 1949. Selective mating in *Drosophila melanogaster*. *Genetics* 34:370–89.
- 1965. Lethal frequency and allelism in DDT-resistant populations and their controls. *Am. Nat.* 99:411–7.
Milkman, R. D. 1978. Modification of heat resistance in *Drosophila* by selection. *Nature (London)* 273:49–50.
Milkman, R. D., and Zeitler, R. R. 1974. Concurrent multiple paternity in natural and laboratory populations of *Drosophila melanogaster*. *Genetics* 78:1191–3.
Misra, R. K., and Reeve, E. C. R. 1964. Clines in body dimensions in populations of *Drosophila subobscura*. *Genet. Res.* 5:240–56.
Moore, J. A., Taylor, C. E., and Moore, B. C. 1979. The *Drosophila* of southern California. I. Colonization after a fire. *Evolution* 33:156–71.
Moray, N., and Connolly, K. 1963. A possible case of genetic assimilation of behaviour. *Nature (London)* 199:358–60.
Morton, R. S. 1966. *Venereal Diseases*. London: Penguin Books.

Moxon, L. N., Holmes, R. S., and Parsons, P. A. Comparative studies of aldehyde oxidase, alcohol dehydrogenase and aldehyde resource utilization among Australian *Drosophila* species. *Comp. Biochem. Physiol.* *71B*:387–95.

Nei, M., Maruyama, T., and Chakraborty, R. 1975. The bottleneck effect and genetic variability in populations. *Evolution 29*:1–10.

Neilson-Jones, W. 1969. *Plant Chimeras*. London: Methuen.

Nevo, E. 1976. Adaptive strategies of genetic systems in constant and varying environments. In *Population Genetics and Ecology* (edited by S. Karlin, and E. Nevo), pp. 141–58. New York: Academic Press.

– 1978. Genetic variation in natural populations: patterns and theory. *Theor. Popul. Biol. 13*:121–77.

Nevo, E., and Yang, S. Y. 1979. Genetic diversity and climatic determinants in tree frogs in Israel. *Oecologia 41*:47–63.

Nix, H. A. 1981. The environment of *Terra Australis*. In *Ecological Biogeography of Australia* (edited by A. Keast), pp. 103–33. The Hague: Junk.

Nix, H. A., and Kalma, J. D. 1972. Climate as a dominant control in the biogeography of northern Australia and New Guinea. In *Bridge and Barrier: The Natural and Cultural History of Torres Strait* (edited by D. Walker), pp. 61–91. Canberra: Australian National University Press.

Novy, J. E. 1978. Operation of a screwworm eradication program. In *The Screwworm Problem* (edited by R. H. Richardson), pp. 19–36. Austin: University of Texas Press.

Oakeshott, J. G., Gibson, J. B., Anderson, P. R., and Champ, A. 1980. Opposing modes of selection on the alcohol dehydrogenase locus in *Drosophila melanogaster*. *Aust. J. Biol. Sci. 33*:105–14.

Ogilvie, D. M., and Stinson, R. H. 1966. Temperature selection in *Peromyscus* and laboratory mice, *Mus musculus*. *J. Mammal. 47*:655–60.

Paik, Y. K., and Sung, K. C. 1969. Behavior of lethals in *Drosophila melanogaster*. *Jap. J. Genet. 44*:180–92.

Parker, E. D., Selander, R. K., Hudson, R. O., and Lester, L. J. 1977. Genetic diversity in colonizing parthenogenetic cockroaches. *Evolution 31*:836–842.

Parry, G. D. 1981. The meanings of r- and K-selection. *Oecologia 48*:260–4.

Parsons, P. A. 1958. Evolution of sex in the flowering plants of South Australia. *Nature (London) 181*:1673–4.

– 1963. Migration as a factor in natural selection. *Genetica 33*:184–206.

– 1965. Assortative mating for a metrical characteristic in *Drosophila*. *Heredity 20*:161–7.

– 1967. *The Genetic Analysis of Behaviour*. London: Methuen.

– 1970. Genetic heterogeneity in natural populations of *Drosophila melanogaster* for ability to withstand desiccation. *Theor. Appl. Genet. 40*:261–6.

– 1971. Extreme-environment heterosis and genetic loads. *Heredity 26*:579–83.

– 1973a. Genetics of resistance to environmental stresses in *Drosophila* populations. *Annu. Rev. Genet. 7*:239–65.

- 1973b. *Behavioural and Ecological Genetics: A Study in Drosophila.* Oxford University Press (Clarendon Press).
- 1974a. Male mating speed as a component of fitness in *Drosophila. Behav. Genet. 4*:395–404.
- 1974b.The behavioral phenotype in mice. *Am. Nat. 108*:377–85.
- 1975a. The comparative evolutionary biology of the sibling species, *Drosophila melanogaster* and *D. simulans. Quart. Rev. Biol. 50*:151–69.
- 1975b. Phototactic responses along a gradient of light intensities for the sibling species *Drosophila melanogaster* and *Drosophila simulans. Behav. Genet. 5*:17–25.
- 1977a. Resistance to cold temperature stress in populations of *D. melanogaster* and *D. simulans. Aust. J. Zool. 25*:693–8.
- 1977b. Genes, behavior, and evolutionary processes: the genus *Drosophila. Adv. Genet. 19*:1–32.
- 1977c. Lek behavior in *Drosophila* (*Hirtodrosophila*) *polypori* Malloch – an Australian rainforest species. *Evolution 31*:223–5.
- 1978a. The genetics of aging in optimal and stressful environments. *Exp. Gerontol. 13*:357–63.
- 1978b. Habitat selection and evolutionary strategies in *Drosophila*: an invited address. *Behav. Genet. 8*:511–26.
- 1979. Larval reactions to possible resources in three *Drosophila* species as indicators of ecological divergence. *Aust. J. Zool. 27*:413–9.
- 1980a. Isofemale strains and evolutionary strategies in natural populations. *Evol. Biol. 13*:175–217.
- 1980b. A widespread Australian endemic *Drosophila* species in New Zealand. *Search 11*:249–50.
- 1980c. Parallel climatic races for tolerances to high temperature-desiccation stress in two *Drosophila* species. *J. Biogeog. 7*:97–101.
- 1980d. Adaptive strategies in natural populations of *Drosophila*: ethanol tolerance, desiccation resistance, and development times in climatically optimal and extreme environments. *Theor. Appl. Genet. 57*:257–66.
- 1980e. Larval responses to environmental ethanol in *Drosophila melanogaster*: variation within and among populations. *Behav. Genet. 10*:183–90.
- 1981a. Evolutionary ecology of Australian *Drosophila*: a species analysis. *Evol. Biol. 14*:297–350.
- 1981b. Longevity of cosmopolitan and native Australian *Drosophila* in ethanol atmospheres. *Aust. J. Zool. 29*:33–9.
- 1982a. Adaptive strategies of colonizing animal species. *Biol. Rev. 57*:117–48.
- 1982b. Acetic acid vapour as a resource and stress in *Drosophila. Aust. J. Zool. 30*:427–33.
- Parsons, P. A., and Hosgood, S. M. W. 1967. Genetic heterogeneity among the founders of laboratory populations of *Drosophila*. I. Scutellar chaetae. *Genetica 38*:328–39.
- Parsons, P. A., MacBean, I. T., and Lee, B. T. O. 1969. Polymorphism in natural populations for genes controlling radioresistance in *Drosophila. Genetics 61*:211–18.

Parsons, P. A., and McDonald, J. 1978. What distinguishes cosmopolitan and endemic *Drosophila* species? *Experientia 34*:1445–6.

Parsons, P. A., and Spence, G. E. 1981a. Longevity, resource utilization and larval preferences in *Drosophila*: inter- and intraspecific variation. *Aust. J. Zool. 29*:671–8.

– 1981b. Acetaldehyde: a low-concentration resource and larval attractant in three *Drosophila* species. *Experientia 37*:576–7.

Parsons, P. A., and Stanley, S. M. 1981. Domesticated and widespread species. In *The Genetics and Biology of Drosophila* (edited by M. Ashburner, H. L. Carson, and J. N. Thompson, Jr.), Vol. 3a, pp. 349–93. New York: Academic Press.

Parsons, P. A., Stanley, S. M., and Spence, G. E. 1979. Environmental ethanol at low concentrations: longevity and development in the sibling species *Drosophila melanogaster* and *D. simulans. Aust. J. Zool. 27*:747–54.

Parsons, P. A., and White, N. G. 1976. Variability of anthropometric traits in Australian Aboriginals and adjacent populations: its bearing on the biological origin of the Australians. In *The Origin of the Australians* (edited by R. L. Kirk, and A. G. Thorne), pp. 227–43. Canberra: Australian Institute of Aboriginal Studies.

Partridge, L. 1978. Habitat selection. In *Behavioural Ecology: An Evolutionary Approach* (edited by J. R. Krebs and N. B. Davies), pp. 351–76. Oxford: Blackwell Scientific Publications.

– 1980. Mate choice increases a component of offspring fitness in fruit flies. *Nature (London) 283*:290–1.

– 1981. Increased preferences for familiar foods in small mammals. *Anim. Behav. 29*:211–6.

Patterson, J. T., and Stone, W. S. 1952. Evolution in the genus *Drosophila*. New York: Macmillan.

Payne, F. 1918. An experiment to test the nature of the variation on which selection acts. *Indiana Univ. Stud. 5(36)*:1–45.

Pennycuik, P., and Fraser, A. S. 1964. Variation of scutellar bristles in *Drosophila*. II. Effects of temperature. *Aust. J. Biol. Sci. 17*:764–70.

Piazza, A., Menozzi, P., and Cavalli-Sforza, L. L. 1981. Synthetic gene frequency maps of man and selective effects of climate. *Proc. Natl. Acad. Sci. USA 78*:2638–42.

Pipkin, S. B., Rhodes, C., and Williams, N. 1973. Influence of temperature on *Drosophila* alcohol dehydrogenase polymorphism. *J. Hered. 64*:181–5.

Powell, J. R. 1971. Genetic polymorphisms in varied environments. *Science 174*:1035–6.

Powell, J. R., Dobzhansky. Th., Hook, J. E., and Wistrand, H. E. 1976. Genetics of natural populations. XLIII. Further studies on rates of dispersal of *Drosophila pseudoobscura* and its relatives. *Genetics 82*:493–506.

Powell, J. R., and Taylor, C. E. 1979. Genetic variation in ecologically diverse environments. *Am. Sci. 67*:590–6.

Powell, J. R., and Wistrand, H. 1978. The effect of heterogeneous environments and a competitor on genetic variation in *Drosophila*. *Am. Nat. 112*:935–47.

Prakash, S. 1973a. Patterns of gene variation in central and marginal populations of *Drosophila robusta*. *Genetics 75*:347–69.
– 1973b. Low gene variation in *Drosophila busckii*. *Genetics 75*:571–6.
– 1977a. Further studies on gene polymorphism in the mainbody and geographically isolated populations of *Drosophila pseudoobscura*. *Genetics 85*:713–9.
– 1977b. Genetic divergence in closely related sibling species *Drosophila pseudoobscura, Drosophila persimilis* and *Drosophila miranda*. *Evolution 31*:14–23.
Prakash, S., Lewontin, R. C., and Hubby, J. L. 1969. A molecular approach to the study of genic heterozygosity in natural populations. IV. Patterns of genic variation in central, marginal and isolated populations of *Drosophila pseudoobscura*. *Genetics 61*:841–58.
Prevosti, A. 1955. Geographic variability in quantitative traits in populations of *D. subobscura*. *Cold Spring Harbor Symp. Quant. Biol. 20*:294–9.
Price, P. W. 1974. Strategies for egg production. *Evolution 28*:76–84.
– 1977. General concepts on the evolutionary biology of parasites. *Evolution 31*:405–20.
– 1980. *Evolutionary Biology of Parasites*, pp. 1–237. Princeton, N.J.: Princeton University Press.
Price, S. C., and Jain, S. K. 1981. Are inbreeders better colonizers? *Oecologia 49*:283–6.
Prince, G. J. 1976. Laboratory biology of *Phaenocarpa persimilis* Papp (Braconidae: Alysiinae), a parasitoid of *Drosophila*. *Aust. J. Zool. 24*:249–64.
Prince, G. J., and Parsons, P. A. 1977. Adaptive behaviour of *Drosophila* adults in relation to temperature and humidity. *Aust. J. Zool. 25*:285–90.
– 1980. Resource utilization specificity in three cosmopolitan *Drosophila* species. *J. Nat. Hist. 14*:559–65.
Prout, T. 1980. Some relationships between density-independent selection and density-dependent population growth. *Evol. Biol. 13*:1–68.
Pruzan, A., and Bush, G. 1977. Genotypic differences in larval olfactory discrimination in two *Drosophila melanogaster* strains. *Behav. Genet. 7*:457–64.
Quinn, W. G., and Dudai, Y. 1976. Memory phases in *Drosophila*. *Nature (London) 262*:576–7.
Quinn, W. G., Harris, W. A., and Benzer, S. 1974. Conditioned behavior in *Drosophila melanogaster*. *Proc. Natl. Acad. Sci. USA 71*:708–12.
Reed, M. E. 1938. The olfactory reactions of *Drosophila melanogaster* to the products of fermenting banana. *Physiol. Zool. 11*:317–25.
Remington, C. L. 1968. The population genetics of insect introduction. *Annu. Rev. Entomol. 13*:415–26.
Rendel, J. M., and Sheldon, B. L. 1960. Selection for canalization of the *scute* phenotype in *Drosophila melanogaster*. *Aust. J. Biol. Sci. 13*:36–47.
Richardson, R. H. 1974. Effects of dispersal, habitat selection and competition on a speciation pattern of *Drosophila* endemic to Hawaii. In *Genetic Mechanisms of Speciation in Insects* (edited by M. J. D. White), pp. 140–64. Sydney: Australia and New Zealand Book Co.

- (ed.) 1978. *The Screwworm Problem*: *Evolution of Resistance to Biological Control*. Austin: University of Texas Press.

Robson, M. K., and Parsons, P. A. 1967. Fingerprint studies on four central Australian Aboriginal tribes. *Archaeol. Phys. Anthropol. Oceania 2*:69–78.

Rockwell, R. F., Cooke, F., and Harmsen, R. 1975. Photobehavioral differentiation in natural populations of *Drosophila pseudoobscura* and *D. persimilis*. *Behav. Genet. 5*:189–202.

Rockwell, R. F., and Seiger, M. B. 1973a. A comparative study of photoresponse in *Drosophila pseudoobscura* and *Drosophila persimilis*. *Behav. Genet. 3*:163–74.

- 1973b. Phototaxis in *Drosophila*: a critical evaluation. *Am. Sci. 61*:339–45.

Rose, M. R., and Charlesworth, B. 1981. Genetics of life history in *Drosophila melanogaster*. I. Sib analysis of adult females. *Genetics 97*:173–86.

Rosenzweig, M. L. 1975. On continental steady states of species diversity. In *Ecology and Evolution of Communities* (edited by M. L. Cody and J. M. Diamond), pp. 121–40. Cambridge, Mass.: Harvard University Press (Belknap Press).

- 1981. A theory of habitat selection. *Ecology 62*:327–35.

Roughgarden, J. 1971. Density-dependent natural selection. *Ecology 52*:453–68.

Sabath, M. D. 1981. Geographical patterns of genetic variability in introduced Australian populations of the marine toad, *Bufo marinus*: Sorbitol dehydrogenase, *Sdh. Biochem. Genetics 19*:347–53.

Sabath, M. D., Boughton, W. C., and Easteal, S. 1981. Expansion of the range of the introduced toad *Bufo marinus* in Australia from 1935 to 1974. *Copeia 3*:676–80.

Safriel, U. N., and Ritte, U. 1980. Criteria for the identification of potential colonizers. *Biol. J. Linn. Soc. 13*:287–97.

Salt, G. 1970. *The Cellular Defence Reaction of Insects*. Cambridge University Press.

Saul, S. H., Sinsko, M. J., Grimstad, P. R., and Craig, G. B., Jr. 1978. Population genetics of the mosquito *Aedes triseriatus*: genetic-ecological correlation at an esterase locus. *Am. Nat. 112*:333–9.

Schaal, B. A., and Levin, D. A. 1976. The demographic genetics of *Liatris cylindracea* Michx. (Compositae). *Am. Nat. 110*:191–206.

Schwartz, D., and Laughner, W. J. 1969. A molecular basis for heterosis. *Science 166*:626–7.

Selander, R. K., and Hudson, R. O. 1976. Animal population structure under close inbreeding: the land snail *Rumina* in southern France. *Am. Nat. 110*:695–718.

Selander, R. K., and Kaufman, D. W. 1973. Self-fertilization and genetic population structure in a colonizing land snail. *Proc. Natl. Acad. Sci. USA 70*:1186–90.

Sewell, D., Burnet, B., and Connolly, K. 1975. Genetic analysis of larval feeding behaviour in *Drosophila melanogaster*. *Genet. Res. 24*:163–73.

Shapiro, A. M. 1976. Seasonal polyphenism. *Evol. Biol. 9*:259–333.

Shorey, H. H. 1977. Interaction of insects with their chemical environment. In *Chemical Control of Insect Behavior: Theory and Application* (edited by H. H. Shorey, and J. J. McKelvey), pp. 1–5. New York: Wiley.

Siegel, P. B. 1979. Behavior-genetics in chickens: a review. *World's Poult. Sci. J. 35*:9–19.

Simmons, R. T., Tindale, N. B., and Birdsell, J. B. 1962. A blood group genetical survey in Australian Aborigines of Bentinck, Mornington and Forsyth Islands, Gulf of Carpentaria. *Am. J. Phys. Anthropol. 20*:303–20.

Simpson, G. G. 1953. *The Major Features of Evolution.* New York: Columbia University Press.

Singh, R. S., Hubby, J. L., and Lewontin, R. C. 1974. Molecular heterosis for heat sensitive enzyme alleles. *Proc. Natl. Acad. Sci. USA 71*:1808–10.

Skellam, J. G. 1951. Random dispersal in theoretical populations. *Biometrika 38*:196–218.

Slatkin, M. 1981. Populational heritability. *Evolution 35*:859–71.

Sokoloff, A. 1965. Geographic variation of quantitative characters in populations of *Drosophila pseudoobscura. Evolution 19*:300–10.

– 1966. Morphological variation in natural and experimental populations of *Drosophila pseudoobscura* and *Drosophila persimilis. Evolution 20*:49–71.

Sonneborn, T. M. 1957. Breeding systems, reproductive methods, and species problems in Protozoa. In *The Species Problem* (edited by E. Mayr), Washington D.C.: American Association for the Advancement of Science.

Soulé, M. 1973. The epistasis cycle: a theory of marginal populations. *Annu. Rev. Ecol. Syst. 4*:165–87.

– 1979. Heterozygosity and developmental stability: another look. *Evolution 33*:396–401.

Southwood, T. R. E. 1977. Habitat, the templet for ecological strategies? *J. Anim. Ecol. 46*:337–65.

Spassky, B., and Dobzhansky, Th. 1967. Responses of various strains of *Drosophila pseudoobscura* and *Drosophila persimilis* to light and gravity. *Am. Nat. 101*:59–63.

Spatz, H. C., Emmens, A., and Reichart, H. 1974. Associative learning of *Drosophila melanogaster. Nature (London) 248*:359–61.

Spiess, E. B. 1970. Mating propensity and its genetic basis in *Drosophila*. In *Essays in Evolution and Genetics in Honor of Theodosius Dobzhansky* (edited by M. K. Hecht and W. C. Steere), pp. 315–79. New York: Appleton-Century-Crofts.

Spieth, H. T. 1952. Mating behavior within the genus *Drosophila* (Diptera). *Bull. Am. Mus. Nat. Hist. 99*: 401–74.

– 1974. Courtship behavior in *Drosophila. Annu. Rev. Entomol. 19*:385–405.

Stalker, H. D. 1976. Chromosome studies in wild populations of *D. melanogaster. Genetics 82*:323–47.

– 1980. Chromosome studies in wild populations of *Drosophila melanogaster*. II. Relationship of inversion frequencies to latitude, season, wing-loading and flight activity. *Genetics 95*:211–23.

Stalker, H. D., and Carson, H. L. 1947. Morphological variation in natural populations of *Drosophila robusta* Sturtevant. *Evolution 1*:237–48.

– 1948. An altitudinal transect of *Drosophila robusta* Sturtevant. *Evolution 2*:295–305.

– 1949. Seasonal variation in the morphology of *Drosophila robusta* Sturtevant. *Evolution 3*:330–43.

Stanley, S. M. 1979. *Macroevolution: Pattern and Process*. San Francisco: Freeman.

Stanley, S. M., and Parsons, P. A. 1981. The response of the cosmopolitan species, *Drosophila melanogaster* to ecological gradients. *Proc. Ecol. Soc. Aust. 11*:121–30.

Stanley, S. M., Parsons, P. A., Spence, G. E., and Weber, L. 1980. Resistance of species of the *Drosophila melanogaster* subgroup to environmental extremes. *Aust. J. Zool. 28*:413–21.

Starmer, W. T. 1981. A comparison of *Drosophila* habitats according to the physiological attributes of the associated yeast communities. *Evolution 35*:38–52.

Starmer, W. T., Heed, W. B., and Rockwood-Sluss, E. S. 1977. Extension of longevity in *Drosophila mojavensis* by environmental ethanol: Differences between subraces. *Proc. Natl. Acad. Sci. USA 74*:387–91.

Stearns, S. C. 1976. Life history tactics: a review of the ideas. *Quart. Rev. Biol. 51*:3–47.

– 1977. The evolution of life history traits: A critique of the theory and a review of the data. *Annu. Rev. Ecol. Syst. 8*:145–71.

Stearns, S. C., and Sage, R. D. 1980. Maladaptation in a marginal population of the mosquito fish, *Gambusia affinis*. *Evolution 34*:65–75.

Stebbins, G. L. 1979. Fifty years of plant evolution. In *Topics in Plant Population Biology* (edited by O. T. Solbrig, S. Jain, G. B. Johnson, and P. H. Raven), pp. 18–41. New York: Columbia University Press.

Stickel, L. F. 1979. Population ecology of house mice in unstable habitats. *J. Anim. Ecol. 48*:871–87.

Strong, D. R., and Levin, D. A. 1979. Species richness of plant parasites and growth form of their hosts. *Am. Nat. 114*:1–22.

Sturtevant, A. H. 1915. Experiments on sex recognition and the problem of sexual selection in *Drosophila*. *J. Anim. Behav. 5*:351–66.

Tabachnick, W. J., and Powell, J. R. 1977. Adaptive flexibility of 'marginal' versus 'central' populations of *Drosophila willistoni*. *Evolution 31*:692–4.

Tallamy, D. W., and Denno, R. F. 1981. Alternative life history patterns in risky environments: an example from lacebugs. In *Insect Life History Patterns: Habitat and Geographic Variation* (edited by R. F. Denno and H. Dingle), pp. 129–47. New York: Springer-Verlag.

Tantawy, A. O., and Mallah, G. S. 1961. Studies on natural populations of *Drosophila* I. Heat resistance and geographical variation in *D. melanogaster* and *D. simulans*. *Evolution 15*:1–14.

Tauber, M. J., and Tauber, C. A. 1978. Evolution of phenological strategies in insects: a comparative approach with eco-physiological and genetic considerations. In *Proceedings in Life Sciences. Evolution of Insect Migration and Diapause* (edited by H. Dingle), pp. 53–71. New York: Springer-Verlag.

Taylor, C. E., and Condra, C. 1980. *r*- and *K*-selection in *Drosophila pseudoobscura*. *Evolution 34*:1183–93.

Taylor, C. E., and Gorman, G. C. 1975. Population genetics of a 'colonising' lizard: natural selection for allozyme morphs in *Anolis grahami*. *Heredity 35*:241–7.

Taylor, C. E. and Powell, J. R. 1977. Microgeographic differentiation of chromosomal and enzyme polymorphisms in *Drosophila persimilis*. *Genetics 85*:681–95.

Taylor, L. R., and Taylor, R. A. J. 1977. Aggregation, migration and population mechanics. *Nature (London) 265*:415–21.

Templeton, A. R. 1980a. The theory of speciation via the founder principle. *Genetics 94*:1011–38.

– 1980b. Modes of speciation and inferences based upon genetic distances. *Evolution 34*:719–29.

– 1981. Mechanisms of speciation - a population genetic approach. *Annu. Rev. Ecol. Syst. 12*:23–48.

Templeton, A. R., and Rothman, E. D. 1981. Evolution in fine-grained environments. II. Habitat selection as a homeostatic mechanism. *Theor. Popul. Biol. 19*:326–40.

Thoday, J. M. 1961. Location of polygenes. *Nature (London) 191*:368–70.

– 1964. Genetics and the integration of reproductive systems. In *Insect Reproduction* (edited by K. C. Highnam), Royal Ent. Soc. Symp. 2, pp. 108–19.

Thoday, J. M., and Boam, T. B. 1961. Regular responses to selection I. Description of responses. *Genet. Res. 2*:161–76.

Thoday, J. M., and Gibson, J. B. 1962. Isolation by disruptive selection. *Nature (London) 193*:1164–6.

Thoday, J. M., Gibson, J. B., and Spickett, S. G. 1964. Regular responses to selection. 2. Recombination and accelerated response. *Genet. Res. 5*:1–19.

Thoday, J. M., and Thompson, J. N., Jr. 1976. The number of segregating genes implied by continuous variation. *Genetica 46*:335–44.

Thompson, J. N., Jr., and Thoday, J. M. 1979. *Quantitative Genetic Variation*, pp. 1–305. New York: Academic Press.

Thomson, G. 1977. The effect of a selected locus on linked neutral loci. *Genetics 85*:753–88.

Thomson, J. A. 1971. Association of karyotype with body weight and resistance to desiccation in *Drosophila pseudoobscura*. *Can. J. Genet. Cytol. 13*:63–9.

Thorne, A. G. 1977. Separation or reconciliation? Biological clues to the development of Australian society. In *Sunda and Sahul: Prehistoric Studies in Southeast Asia, Melanesia and Australia* (edited by J. Allen, J. Golson, and R. Jones), pp. 187–204. New York: Academic Press.

Thorne, A. G., and Wolpoff, M. H. 1981. Regional continuity in Australasian Pleistocene hominid evolution. *Am. J. Phys. Anthropol. 55*:337–49.

Thorpe, W. H. 1945. The evolutionary significance of habitat selection. *J. Anim. Ecol. 14*:67–70.

Throckmorton, L. J. 1975. The phylogeny, ecology, and geography of *Drosophila*. In *Handbook of Genetics* (edited by R. C. King), Vol. 3, pp. 421–69. New York: Plenum Press.

Tindale, N. B. 1962. Some population changes among the Kaiadilt people of Bentinck Island, Queensland. *Rec. S. Aust. Mus. 14*:297–336.

Townsend, J. I. 1952. Genetics of marginal populations of *Drosophila willistoni. Evolution 6*:428–42.

Tracey, J. G., and Webb, L. J. 1975. *Key to the Vegetation of the Humid Tropical Regions of North Queensland*. Brisbane: CSIRO, Long Pocket Laboratories.

Trivers, R. L. 1972. Parental investment and sexual selection. In *Sexual Selection and the Descent of Man* (edited by B. Campbell), pp. 136–79. Chicago: Aldine.

Tucić, N. 1979. Genetic capacity for adaptation to cold resistance at different developmental stages of *Drosophila melanogaster. Evolution 33*:350–8.

Ucko, P. J., and Dimbleby, G. W. (eds.) 1969. *The Domestication and Exploitation of Plants and Animals*. London: Duckworth.

Waddington, C. H. 1965. Introduction to the symposium. In *The Genetics of Colonizing Species* (edited by H. G. Baker, and G. L. Stebbins), pp. 1–6. New York: Academic Press.

Wallace, A. R. 1876. *The Geographical Distribution of Animals*. London: Macmillan.

– 1880. *Island Life*. London: Macmillan.

Wallace, B. 1966a. On the dispersal of *Drosophila. Am. Nat. 100*:551–63.

– 1966b. Distance and the allelism of lethals in a tropical population of *Drosophila melanogaster. Am. Nat. 100*:565–78.

– 1968. *Topics in Population Genetics*. New York: Norton.

– 1970. Observations on the microdispersion of *Drosophila melanogaster*. In *Essays in Evolution and Genetics in Honor of Theodosius Dobzhansky* (edited by M. K. Hecht and W. C. Steere), pp. 381–99. New York: Appleton-Century-Crofts.

– 1975. The biogeography of laboratory islands. *Evolution 29*:622–35.

– 1978. The adaptation of *Drosophila virilis* to life on an artificial crab. *Am. Nat. 112*:971–73.

– 1981. *Basic Population Genetics*. New York: Columbia University Press.

Wasserman, S. S., and Futuyma, D. J. 1981. Evolution of host plant utilization in laboratory populations of the southern cowpea weevil, *Callosobruchus maculatus* Fabricius (Coleoptera: Bruchidae). *Evolution 35*:605–17.

Watanabe, T. K., and Kawanishi, M. 1976. Colonization of *Drosophila simulans* in Japan. *Proc. Jap. Acad. 52*:191–4.

Watts, D. 1971. *Principles of Biogeography*. New York: McGraw-Hill.

Webb, L. J., and Tracey, J. G. 1972. An ecological comparison of vegetation communities on each side of Torres Strait. In *Bridge and Barrier: The Natural and Cultural History of Torres Strait* (edited by D. Walker), pp. 109–29. Canberra: Australian National University Press.

– 1981. The rainforests of northern Australia. In *Australian Vegetation* (edited by R. H. Groves), pp. 67–101. Cambridge University Press.

Wecker, S. C. 1964. Habitat selection. *Sci. Am. 211*(4):109–116.

West, A. S. 1961. Chemical attractants for adult *Drosophila* species. *J. Econ. Entomol. 54*:677–81.

Westerman, J. M., and Parsons, P. A. 1972. Radioresistance and longevity of inbred strains of *Drosophila melanogaster. Int. J. Radiat. Biol. 21*:145–52.

– 1973. Variations in genetic architecture at different doses of γ-radiation as measured by longevity in *Drosophila melanogaster. Can. J. Genet. Cytol. 15*:289–98.

White, E. B., DeBach, P., and Garber, M. J. 1970. Artificial selection for genetic adaptation to temperature extremes in *Aphytis lingnanensis* compere (Hymenoptera: Aphelinidae). *Hilgardia 40*:161–92.

White, M. J. D. 1973. *Animal Cytology and Evolution.* Cambridge University Press.

White, N. G. 1978. A human ecology research project in the Arnhem Land region: An Outline. *Aust. Inst. Aboriginal Stud. Newslett. 9*:39–52.

– 1979a. The use of digital dermatoglyphics in assessing population relationships in Aboriginal Australia. In *Birth Defects: Original Article Series,* Vol. XV, No. 6, pp. 437–54. Washington, D.C.: The National Foundation.

– 1979b. *Tribes, Genes, and Habitats: Genetic Diversity among Aboriginal Populations in the Northern Territory of Australia.* Ph.D. Thesis, La Trobe University, Bundoora, Australia.

White, N. G., and Parsons, P. A. 1973. Genetic and sociocultural differentiation in the Aborigines of Arnhem Land Australia. *Am. J. Phys. Anthropol. 38*:5–14.

– 1976. Population genetic, social, linguistic and topographical relationships in north-eastern Arnhem Land, Australia. *Nature (London) 261*:223–5.

White, N. G., Stanley, S. M., and Parsons, P. A. 1981. Populations, habitats and colonising strategies (in Australia). *Search 12*:251–2.

Whitham, T. G. 1981. Individual trees as heterogeneous environments: adaptation to herbivory and epigenetic noise? In *Insect Life History Patterns: Habitat and Geographic Variation* (edited by R. F. Denno and H. Dingle), pp. 9–27. New York: Springer-Verlag.

Whitham, T. G., and Slobodchikoff, C. N. 1981. Evolution by individuals, plant-herbivore interactions, and mosaics of genetic variability: the adaptive significance of somatic mutations in plants. *Oecologia 49*:287–92.

Whittaker, R. H. 1977. Evolution of species diversity in land communities. In *Evolutionary Biology* (edited by M. K. Hecht, W. C. Steere, and B. Wallace), Vol. 10, pp. 1–67. New York: Plenum Press.

Wiens, J. A. 1977. On competition and variable environments. *Am. Sci. 65*:590–7.

Wigley, T. M. L., and Atkinson, T. C. 1977. Dry years in south-east England since 1698. *Nature (London) 265*:431–4.

Williams, E. E. 1969. The ecology of colonization as seen in the zoogeography of anoline lizards on small islands. *Quart. Rev. Biol. 44*:345–88.

Williams, G. C. 1975. *Sex and Evolution*. Princeton, N.J.: Princeton University Press.

Williamson, P. G. 1981. Palaeontological documentation of speciation in Cenozoic mulluscs from Turkana basin. *Nature (London) 293*:437–43.

Wills, C. 1978. Rank-order selection is capable of maintaining all genetic polymorphisms. *Genetics 89*:403–47.

Wilson, E. O. 1961. The nature of the taxon cycle in the Melanesian ant fauna. *Am. Nat. 95*:169–93.

– 1965. The challenge from related species. In *The Genetics of Colonizing Species* (edited by H. G. Baker and G. L. Stebbins), pp. 7–24. New York: Academic Press.

Wood-Gush, D. G. M. 1972. Strain differences in response to sub-optimal stimuli in the fowl. *Anim. Behav. 20*:72–6.

Wright, S. 1980. Genic and organismic selection. *Evolution 34*:825–43.

Ziegler, J. R. 1976. Evolution of the migration response: emigration by *Tribolium* and the influence of age. *Evolution 30*:579–92.

Ziolo, L. K., and Parsons, P. A. 1982. Ethanol tolerance, alcohol dehydrogenase activity and *Adh* allozymes in *Drosophila melanogaster*. *Genetica 57*:231–7.

Zouros, E., Singh, S. M., and Miles, H. E. 1980. Growth rate in oysters: an overdominant phenotype and its possible explanations. *Evolution 34*:856–67.

Index